LEARNING

SAS

IN THE

COMPUTER LAB

Rebecca J. Elliott
University of Utah

Duxbury Press

An Imprint of Wadsworth Publishing Company

I⟨T⟩P™   An International Thomson Publishing Company

Belmont • Albany • Bonn • Boston • Cincinnati • Detroit • London • Madrid • Melbourne
Mexico City • New York • Paris • San Francisco • Singapore • Tokyo • Toronto • Washington

Software Development Editor: Stan Loll
Editorial Assistant: Claire Masson
Production Editor: Julie Davis
Designer: Ann Butler
Print Buyer: Karen Hunt
Copy Editor: Robert Fiske
Cover: Image House, Inc./Stuart Paterson
Printer: Malloy Lithographing, Inc.

Printed in the United States of America
2 3 4 5 6 7 8 9 10—01 00 99 98 97 96 95

For more information, contact Wadsworth Publishing Company:

Wadsworth Publishing Company
10 Davis Drive
Belmont, California 94002, USA

International Thomson Editores
Campos Eliseos 385, Piso 7
Col. Polanco
11560 México D.F. México

International Thomson Publishing Europe
Berkshire House 168-173
High Holborn
London, WC1V 7AA, England

International Thomson Publishing GmbH
Königswinterer Strasse 418
53227 Bonn, Germany

Thomas Nelson Australia
102 Dodds Street
South Melbourne 3205
Victoria, Australia

International Thomson Publishing Asia
221 Henderson Road
#05-10 Henderson Building
Singapore 0315

Nelson Canada
1120 Birchmount Road
th, Ontario
K 5G4

International Thomson Publishing Japan
Hirakawacho Kyowa Building, 3F
2-2-1 Hirakawacho
Chiyoda-ku, Tokyo 102, Japan

Congress Cataloging-in-Publication Data

ecca J.
earning SAS in the Computer Lab/Rebecca J. Elliott
        p.   cm.
cludes index.
BN: 0-534-23442-9
SAS (Computer file) 2. Statistics—Data processing.

QA276.4.E44   1995
519.5'078—dc20   94-27582

Preface

This manual was developed to integrate computing into a beginning statistics class. Newly published statistics texts contain minimal information about using computer packages to analyze data. Statistics instructors would like to have students analyze real data. The problem has been that it is difficult and time consuming to create a computer component to an existing class. Information in package manuals rarely offers information about the software in a form that is easy for a beginner to learn. *Learning SAS in the Computer Lab* is a structured way for teachers of statistics to include computing in their classes.

Although many different statistics packages are on the market, the SAS programming language is widely held to be the standard. Few packages offer the capability to analyze all types of statistics problems and the exceptional database management techniques that SAS offers. SAS can be used on many different operating environments, using the same code on all of them. Although SAS is not as user friendly as some packages, students will be able to learn the basic data analysis techniques and finish the course with an appreciation of computing and analysis. And they will have learned a valuable skill for future work with statistics.

What this manual is and isn't This manual is not a standalone course in statistics. It does not attempt to teach statistics or explain statistical concepts in any depth. It is intended to be used in conjunction with a statistics course where an instructor explains statistical concepts and leads the discussion on "what it all means." In addition, this manual is not a SAS reference book. It does not attempt to cover all the procedures, options in procedures, or database management capabilities in SAS.

This manual provides the means for students to do some computing in a statistics course and acquire an appreciation for data analysis. They will learn about the SAS programming language. They will learn more about computers and computing than from some of the "easier" packages. They will learn enough about SAS "basics" to find their way around the many SAS reference manuals.

The modules There are 21 modules in this manual. The statistical concepts range from simple descriptive statistics to ANOVA, regression, and analysis of covariance. Each module contains a discussion of new SAS commands and procedures. Examples include SAS programs with discussions of any new statements used in the example program and the output generated. At the end of each module is a list of the key ideas covered.

Because instructors use different texts and like to cover topics in different orders, the organization of the manual was designed to be flexible. The SAS commands necessary to get started with any analysis are contained in the first two modules. The rest of the modules can be used in a different order. At the beginning of each module, there is a summary of which previous modules are needed to understand the new SAS concepts.

There are a lot of choices for emphasizing different points in statistics. Some modules can be combined. Module 3, containing information on data management, was added at the request of the reviewers and could easily be skipped. Module 4 contains information on SAS functions and is included mainly as a reference. It also could easily be skipped. Modules 5, 6, 7, and 9 deal with simple descriptive statistics and plots. Module 8, "Generating Random Observations," is a nice way for students to see firsthand how the shape of the distribution of a sample mean differs from the shape of the underlying distribution. Modules 10 and 11 address one-sample and two-sample tests. Modules 12 and 13 cover analysis of variance. Module 14 looks at correlations. Modules 15 through 18 cover simple and multiple linear regression. Module 19 is chi-square tests, 20 is nonparametric ANOVA, and 21 covers analysis of covariance.

The problems Each module contains several problems. The students must write a SAS program to analyze the data, which is available on a computer disk. The data sets are real. Some have been modified slightly to meet the needs of a beginning statistics class. Others are subsets of larger data sets.

The data sets are described in the Appendix. This description includes some background on the data, the number of observations, which modules refer to the data set, the variables and how they appear in the file, and the first three lines of the file so that the students can "see" what the data looks like.

Instructors may wish to use the data sets in ways other than those presented. They can assign different analyses or ask different questions about the data. They can choose to emphasize specific points about data analysis. They can assign problems from text books or have students use data sets not provided with this manual. The manual teaches the students how to use SAS to do particular types of analyses that can also be applied to other sets of data.

Student background This manual can be used for any beginning statistics course. The students do not need to know calculus or have had any previous computing experience.

In order to write and submit SAS programs, the students will need some additional information that is site specific. They will need to know how to use an editor to create, modify, and save files. They will need to know how to have SAS access separate files that contain data. They will also need to know how to submit SAS jobs and how to print the results and other files. Instructors may also want the students to know how to send and receive e-mail messages. Each location offers different editors, has different commands for saving and printing files, and has different ways to access the computer system. Students would find a handout with this information to be valuable. One or two lab sessions dedicated to sharing this information and giving the students time to practice would also be valuable.

The original intent of this manual was for it to be used in a computer lab setting with a TA or instructor available, but it is possible for someone to learn SAS through self-study with minimal help from computer technicians. Several students have completed the lab requirement at the University of Utah this way because of course scheduling conflicts.

At the University of Utah This manual is the culmination of five years of work to incorporate a computing component into a five-credit hour, one-quarter calculus-based statistics course that covered beginning probability, confidence intervals, hypothesis testing, ANOVA, and regression. Initially, students were asked to do a few homework assignments in the computer lab after minimal instruction on SAS code during regular lecture periods. Then, a TA was assigned to present SAS information one day a week. Handout materials contained information on the computer system and some SAS procedures. Eventually, the one-quarter course became a two-quarter course, one of four credits and one of three credits, and each having a one-credit two-hour required lab. The Statistics Committee at the University of Utah commissioned me to write the manuals that would accompany these courses. This manual is the extension of those two manuals for a broader audience.

At the University of Utah, this manual is used in a two-quarter two-hour computer lab of 20 to 25 students, where a TA "lectures" for 10 to 15 minutes. The TA is available for one-on-one help throughout the remainder of the lab. In the first lab session, students get a user-id, logon to the system, and practice sending and receiving e-mail messages. In the second session, they learn how to use an editor. In subsequent labs, students are assigned some or all of the problems at the end of each module and are required to submit their programs, some of the output, and a writeup that answers questions found in the problems and supplemental questions written by the TA. These assignments are worth 10 to 20 points, where half the points are for the written part. The course grade consists of the homework and a final comprehensive exam.

The TAs are graduate students in the statistics program and had no previous SAS experience. The learned SAS by reading this manual. Occasionally, they needed to ask me for help and to refer to SAS manuals.

The source of the SAS information I have been programming with SAS for 12 years. The information presented in this manual is based on my own experience with SAS. I also referred to the several manuals that come with SAS Version 6 whenever I needed specific details.

Those needing additional information about SAS could reference *SAS Application Programming* by Frank DiIorio or the many SAS manuals.

Supplements The data sets described in the Appendix are available on a computer disk, free to those who adopt this manual. A solutions manual containing SAS programs for each of the problems is also available.

Acknowledgments I would like to thank the many individuals who helped me with this project. I send special thanks to the University of Utah Statistics Committee, which originally commissioned me to write the manuals that have become this book.

I am grateful to the many students who took time to comment on what they liked and didn't like about the original manuals and the labs. I also thank the reviewers— Gordon A. Allen, Miami University, Cindy R. Ford, Southern Methodist University, Julia A. Norton, California State University at Hayward, Charles J. Sommer, SUNY College at Brockport, and Timothy A Wittig, Mississippi State University—for their helpful suggestions and encouraging criticisms. I incorporated many of these ideas in the final version of the manual.

The lab TAs found many mistakes in the original manuals and offered their comments and suggestions on wording, content, and assignments. I am particularly grateful to TAs Larry Cook, Jeff Miyatake, and Monika Serbinowska; also Tina Ma and Laura Smithies.

I also want to thank SAS Institute for providing the SAS System for Microsoft Windows for me to use to check the problems. Having the software on the same computer as this manuscript made my job much easier.

And finally, I want to thank my husband, Professor J. David Mason, for his support and belief in this project and for the role he plays on the Statistics Committee to continually improve the statistics classes offered at the University of Utah.

Rebecca J. Elliott

TABLE OF CONTENTS

Introduction

Everyone uses statistics. Statistics provide information about people, processes, events, and ideas. They are used to help make business decisions and as guides for future actions. Whatever your field of study, at the least you will be asked to understand and interpret statistics. More and more often, you will be asked to compute statistics.

Statistical analysis almost always requires the use of computers because of the large number of observations in a data set and because the calculations are tedious and often complicated. Because of the importance and benefits of computers in our society, instructors of statistics want to incorporate computing in their courses. *Learning SAS in the Computer Lab* is a way for you to analyze data, letting the computer do all the hard work. The SAS programming language is the standard by which all the other packages are measured. It has excellent data management tools, which are needed because data are not always in the best form to be analyzed. Knowing some SAS when you are job hunting could be a real benefit!

This manual is the outgrowth of materials developed to integrate a computing component into a beginning statistics course. The intent is not to teach statistics but to provide a way for you to learn how to do statistics on the computer. Because this manual does not teach statistical concepts, it must be a supplement to a statistics course or used by someone who has a statistics background and who wants to learn SAS. People who know little about statistics and think this manual will help them choose the correct analysis for their data will be disappointed. The manual assumes the reader knows statistics or is currently in a statistics course.

Although this manual was designed to be used in a computer lab with an instructor, it is possible for you to learn on your own with a little bit of help from a computer person to get you started on the computer system. You need to know how to use an editor, how to save and print files, how to submit batch SAS jobs, and how to have SAS read data from computer files. This information can be incorporated into labs or learned through knowledgeable sources.

The 21 modules cover descriptive statistics, t-tests, ANOVA, regression, chi-square tests, and analysis of covariance—more topics than are usually covered in just one beginning statistics course. The modules contain explanations of SAS code, example programs, and a discussion of what to look for on the output SAS generates. The modules can be used in many different orders. At the beginning of each one is a list of modules that should have been covered first in order for you to understand the SAS code used.

The problems consist of writing SAS programs to analyze real data sets, which are provided on a floppy disk free to those who adopt this manual for a course.

Once you learn some of the basics, you may want to browse through the many manuals SAS offers to discover more of SAS's capabilities. If there's something you need to do to data, there is probably a way to do it with SAS.

The Basics

Using SAS to analyze data requires an understanding of the statements and keywords that make up a SAS program. Although SAS programs can be very complicated and can do very complicated things, it is best to begin with the basics of a simple SAS program. In this module, you will learn how to write a program that reads and prints a set of data.

DATA STEPS AND PROC STEPS

SAS is organized into steps, which are like paragraphs. There are two types of steps: DATA steps, which put data in a form that the SAS program can use, and PROC steps, which use procedures to do something to the data, such as sorting it, analyzing it, or printing it. Each SAS step or paragraph must start with DATA or PROC to let SAS know what kind of step it is.

SAS steps consist of statements, which are like sentences in a paragraph. A semicolon (;) is required to denote the end of a statement, just like a period is required at the end of a sentence. SAS statements consist of keywords that have special meanings in SAS and other words that the programmer adds, such as the names of variables.

Spacing does not matter: You can put as many spaces between the words and keywords of statements as you like, and you can have as many blank lines in a program as you like. The semicolon indicates the end of statements. Many programmers use a consistent layout pattern to make their SAS programs easier to read and, thus, easier to debug or correct. You may find it useful to use the program layout shown in the examples of this manual, which clearly delineates between DATA steps and PROC steps.

THE DATA STEP

In a simple SAS program, one begins by setting up the DATA step. The DATA step consists of statements that create a data set, which SAS can then analyze in subsequent PROC steps.

The DATA statement names the data set. A data set can have any name you like as long as it starts with a letter and has no more than eight characters of numbers, letters, underscores, or other special characters. Valid names for data sets include F4, PESON_1, NEW, and X5Z8.

INPUT is the keyword that defines the names of the variables in the data set. You can use any name for a variable as long as it begins with a letter and has at most eight characters. One of the nice things about SAS is that you can name variables something that makes sense and that will be easy to remember. For instance, you can name two variables HEIGHT and WEIGHT instead of X and Y (although you could name them X and Y if you wanted).

The CARDS statement signals the beginning of the lines of data. It is a holdover from the days when programmers wrote computer programs on punch cards. Spacing in data lines *does* matter. A semicolon must be on the line following the last line of data. You will learn more about inputting data later in this module.

RUN is an optional last statement in DATA steps and PROC steps when you use batch processing for SAS programs. Batch processing means that you write the entire program and then submit it to the SAS processor, which performs the actions requested by each statement. In interactive SAS sessions, statements are evaluated after the RUN statements, which are required. The RUN statement makes the LOG file (described later in this module) easier to read. It basically serves as a "last sentence" in the DATA or PROC step.

General Form of Simple Data Step

```
data data set;
    input variables;
cards;
the lines of data
;
run;
```

PROC PRINT

This procedure tells SAS to print out certain variables in the data set. When it prints, it will provide the variable names and the values that SAS reads.

The keyword VAR is short for variable list. You list the variables you want to print after VAR in the order you want them printed. If you omit this statement, PROC PRINT will print all the variables in the data set in the order in which they were read. It is advisable to print out data sets (unless there are many observations) to make sure SAS read the data the way you intended.

The BY statement can be used in all SAS procedures and will be discussed in Module 3.

PROC PRINT General Form

```
proc print data=data set;
    by variables;
```

```
   var variables;
run;
```

EXAMPLE 1.1

An analyst has collected data on four variables (height in inches, weight in pounds, age in years, and name) for three people. She wants to read the data into the computer and then print it out. Here is a simple SAS program to do that.

SAS Program

```
data one;
   input height weight name $ age ;
cards;
65      150      Chris      50
60      125      Kelly      35
68      180      Leslie     29
;
run;

proc print data=one;
   var name age weight height;
run;
```

New Statements

data one;   ONE is the name of this data set. Notice the semicolon that signals the end of the DATA statement.

input height weight name $ age;

HEIGHT is the name of the first variable, WEIGHT is the name of the second variable, NAME is the name of the third variable, and AGE is the name of the last variable. The dollar sign $ indicates that the variable NAME is a character variable (more on this later). Again, notice the semicolon at the end of this statement.

```
cards;
65    150    Chris    50
60    125    Kelly    35
68    180    Leslie   29
;
```

There are three lines of data in this program. Data are grouped by case or observation. The INPUT statement said there were four variables, HEIGHT, WEIGHT, NAME, and AGE for each observation (person). Thus, for the first observation, HEIGHT=65, WEIGHT=150, NAME=Chris, and AGE=50; for the second observation,

HEIGHT=60, WEIGHT=125, NAME=Kelly, and AGE=35; and for the third, HEIGHT=68, WEIGHT=180, NAME=Leslie, and AGE=29. When you run a SAS program, it will tell you how many observations are in the data set and how many variables there are. Here, there are three observations and four variables. Notice that there is a semicolon on the line following the last line of data and there are no semicolons at the end of the individual data lines. A semicolon must be on the line following the last line of data. You will learn more about inputting data later in this module.

```
run;
```
This is the "last sentence" in the DATA step.

```
proc print data=one;
var name age weight height;
run;
```
DATA=ONE tells PROC PRINT to use the SAS data set named ONE.

THE INPUT STATEMENT

The INPUT statement names the variables in your data set and tells SAS where the values of the variables can be found. The INPUT statement has many forms to read many different kinds of data formats, such as dates, binary data, hexidecimal data, and packed data. Don't be concerned if you don't know what these terms mean. This module will look at list input, column input, and formatted input to read numeric and character data.

Variables can be either numeric or character (also called alphanumeric). We use numeric data for doing calculations, such as computing the average height of a group of people, finding the total cost of telephone service for one year, or determining the highest grade point average.

SAS assumes that variables are numeric unless you designate that the variables are character. If you forget to designate that a variable is character, SAS will give you an error message that says it found invalid data for a numeric variable. Sometimes we use numbers to code data, but we will never do arithmetic with those values. For instance, the variable home ownership could be a 1 if one owns a home and 2 if one rents an apartment. It would not make any sense to do arithmetic with these values of 1 and 2. When this happens, you could declare the variable home ownership to be character. The way you designate character depends on the type of INPUT statement, although it always involves the use of a dollar sign ($):

List Input

In Example 1.1, data is read using list input. In list input, the data are in a list: The values are separated by spaces. The first group of letters or numbers that SAS finds is assigned to the first variable on the INPUT statement, the next group of letters or numbers that SAS finds is assigned to the next variable on the INPUT statement, and so on. It is the spaces between the groups of letters and numbers in the data set that separate the different variables.

You designate that a variable is character by placing a dollar sign ($), surrounded by spaces, after the variable name. This input format works for character data as long as the data values are eight or fewer characters: If values are longer than eight characters, they will be truncated to eight characters.

Column Input

In order to use column input, your data must be in the same columns on every line. With column input, you specify after the variable name the column positions (out of 80 possible positions) where it can be found. If a variable is character, a dollar sign ($) would come before the column designation. By pointing to the location of variables, you can skip columns to omit some data, you can read the variables in any order, you can read part of a variable, and you can have embedded blanks in character data.

EXAMPLE 1.2

Becky has a data file that contains information about her siblings: name, current age, birth date, and favorite sport. The data start on the first line of the file. The line with the numbers and dashes is to help you see which columns the data are in. (It is not part of the file.)

```
----+----1----+----2----+----3----+----4
Ronald   40  Dec  3 1954   golf
Michael  37  Jul  4 1957   fishing
Laurel   33  Jun 23 1961   softball
```

Here are several possible input statements to read this data:

```
input name $ 1-7 age 10-11 birthdat $ 14-24
     sport $ 28-35;

input name $ 1-7 birthmon $ 14-16 birthyr 23-24
     age 10-11 sport $ 28-35;

input sport $ 28-35 age 10-11 birthday $ 14-19
     name $ 1-7 birthdat $ 14-24;
```

As you can see, it is possible to read the variables in any order, to skip columns, and to reread columns. Notice that the character variable BIRTHDAT contains embedded blanks. It would not be possible to accurately read the values of this variable using list input.

Formatted Input

Imagine that there is a pointer that moves along the data line. In a way, column input used a pointer because you specified which columns to read for each variable. Formatted input allows you to indicate the beginning column of a variable and then specify the length of the variable. This module will focus on two pointer controls and two other character formats.

EXAMPLE 1.3

Let's assume Becky has added more information to her siblings file. She has a second line containing information on year graduated from high school, number of children, and occupation.

```
----+----1----+----2----+----3----+----4
Ronald   40  Dec  3 1954   golf
         1973   2    masonry contractor
Michael  37  Jul  4 1957   fishing
         1975   2    bricklayer
Laurel   33  Jun 23 1961   softball
         1979   0    attorney
```

Here is a formatted input statement to read the data from this example.

```
input @1 name $7. @10 age 2.0 @14 birthdat $char11.
      @28 sport $8. / @9 gradyr 4.0 @16 numchild 1.0
      @20 occupatn $char20.;
```

New Formats

@1, @10, @14, @28

The @ symbol says go to the column number that follows to read the variable. With this "pointing device," you can read variables in any order, skip variables, and reread columns of data just as with column input.

/

This tells the pointer to go to the next line. Once you go to the next line, you cannot move back to the previous line.

2.0, 4.0, 1.0 This is a numeric format $x.y$, indicating that the variable has length x with y decimal places. If the decimal place is indicated in the value, the $.y$ part is ignored. For instance, you may have entered in your file the number 23.46 as 2346, omitting the decimal point. The format would be 4.2, indicating that the value of the variable is in the next four columns and that there are really two decimal places. SAS will recognize that value as 23.46 for any analysis you do.

$7., $8.   This is the $w. format. It indicates that the variable is character and has length w. Notice the period that follows the length w. It is part of the format.

$char11., $char20.

This is the $charw. format. It indicates that the variable is character and has length w. Notice the period that follows the length w. It is part of the format.

COMPARISON OF CHARACTER FORMATS

The three character format ($, $w., and $charw.) differ in how they handle blanks and missing values. Often when collecting data, some of the information is not available or is missing. SAS recognizes a period (.) as missing for numeric data. It usually recognizes blanks as missing for character data.

If you use list input, you must use a . to indicate missing data for numeric or character variables, or the next group of nonblank columns will be assigned to the wrong variable. With list input there cannot be any blanks in character variables since it recognizes a blank as the end of the data value. List input also truncates character values to eight characters.

With the $w. format, the data value can contain embedded blanks, while any leading blank spaces are truncated. It always left-justifies the data value. If you use $w. when the columns contain all blanks, the character variable would be all blanks and, thus, missing.

The $charw. format does not trim leading blanks. It reads a . as a character value and not as missing. It is important whether there are leading blanks or not when you compare two data values. Using an underscore (_) to indicate a blank space, SAS would not find these two character values, each of length five, to be the same: _golf and golf_.

SAS PROGRAM, LOG, AND LIST FILES

In order to write a computer program, you must create a file using an editor. Editors are somewhat like word processors in that they let you add, delete, and move lines; change words; and save what you have done. The editor you use will depend on the computer system you have. After you have written your SAS program, you will save it and then submit it for batch processing. Once SAS is finished carrying out your program commands, you will have three files instead of the one file you started with.

The original file is the SAS program file. This file is the one you create containing all the statements in your program. SAS does not change anything in this file. You may want to save the program files you create for homework problems since most data files are used in subsequent modules.

The second file is the log file, which contains a "log" of the computer's processing of your program. Error messages are found here, as well as informative statements about processing time and warnings that the analysis generated could be incorrect. The log file for Example 1.1 is shown in Figure 1.1. Note that the log file contains your program statements, which are numbered, followed by SAS messages, which begin with "NOTE." SAS messages will begin with "ERROR" if you've made a programming mistake or with "WARNING" if SAS thinks you may be making a mistake.

Figure 1.1   SAS Log File for Example 1.1

```
1                              The SAS System                11:46
Wednesday, April 14, 1993

NOTE: Copyright(c) 1989 by SAS Institute Inc., Cary, NC USA.
NOTE: SAS (r) Proprietary Software Release 6.07  TS110
      Licensed to UNIVERSITY OF UTAH, Site 0005550013.

1           data one;
2              input height weight name $ age ;
3              cards;

NOTE: The data set WORK.ONE has 3 observations and 4 variables.
NOTE: DATA statement used:
      real time           0.172 seconds
      cpu time            0.055 seconds


7           ;
8           run;
9
10          proc print data=one;
11             var name age weight height;
12          run;

NOTE: The PROCEDURE PRINT printed page 1.
NOTE: PROCEDURE PRINT used:
      real time           0.402 seconds
      cpu time            0.055 seconds


NOTE: The SAS System used:
      real time           1.930 seconds
      cpu time            0.590 seconds

NOTE: SAS Institute Inc., SAS Circle, PO Box 8000, Cary, NC 27512-8000
```

The list file contains the actual output that is generated by SAS procedures. If the log file indicates that your program "bombed" (it didn't run successfully), then no list file would be generated. Getting a list file is no guarantee that you got what you wanted. It is always necessary to check to make sure that SAS did what you really wanted it to, not just what you actually (inadvertently) told it to do. See Figure 1.2, the output generated by PROC PRINT in Example 1.1.

Figure 1.2   SAS List File for Example 1.1

```
   The SAS System              11:46 Wednesday, April 14, 1993   1

          OBS      NAME      AGE     WEIGHT     HEIGHT
           1       Chris     50       150        65
           2       Kelly     35       125        60
           3       Leslie    29       180        68
```

PROBLEMS

1.1   Write a SAS program to read and print the following data set. Submit this program.

pH	Time (min)	Temperature
4.5	20	125
4.1	22	133
4.8	18	149
4.0	26	120
5.0	25	120
6.0	21	138

1.2   Write and submit a SAS program to read in the data of Problem 1.1, and print the variables temperature and pH, in that order.

1.3   Write and submit a SAS program to read and print the following data set.

Size	Color	Price (dollars)	Shipping Cost (dollars)
Large	Red	18.97	0.25
Medium	Blue	24.68	1.10
X-Large	Black	29.99	1.75
Small	Orange	15.89	0.90

1.4   Write and submit a SAS program to read the data from Problem 1.3, and print the variables color, size, and price, in that order.

1.5   Write and submit a SAS program to read and print the following data set.

School District	# of Teachers	# of Students
Granite	5,829	200,486
Jordan	12,433	318,992
Davis	2,358	126,331

1.6   Redo Problem 1.1 using column input.

1.7   Redo Problem 1.1 using formatted input.

1.8   Redo Problem 1.3 using column input.

1.9   Redo Problem 1.3 using formatted input.

1.10   Write and submit a program using column input to read the following data about Tom's Tuesday appointments. Print the data.

Time	With	Place	Subject	Length of Meeting
11:00	Sally	Room 30	Personnel review	45 minutes
1:00	Jim	Jim's office	Brake design	30 minutes
3:00	Nancy	Lab	Test results	30 minutes

1.11   Redo Problem 1.10 using formatted input.

YOU SHOULD NOW KNOW

the difference between DATA steps and PROC steps

how to use the CARDS statement to input data

how to use PROC PRINT to get a listing of the data set

how to use list, column, and formatted INPUT statements

how to handle missing data

the difference between character formats

More SAS Basics
MODULES NEEDED 1

SAS is a very powerful data management program. You can create new variables that are based on variables in your data set. The new variables can even be based on complicated criteria. For instance, you may have data that reflects gas mileages for different types of cars. You may be interested in changing miles per gallon into kilometers per liter. Or you may want to create a new variable based on the gas mileage: If the mileage is < 20, then new variable = bad; if the mileage is between 20 and 30, new variable = okay; and if the mileage is > 30, new variable = good.

You may want to put the data in a different order, such as printing all the cars with good mileage first.

It is also possible to use a subset of observations in the data set based on certain criteria. You may be interested in looking at the data for cars made in the United States and excluding those made in Japan.

You may have data that already exists in another file. Why type it into the SAS program? SAS can read the data from outside the SAS program.

In this module, you will learn how to create new variables in SAS, how to subset the data, how to read data from a file, and how to sort the data.

SUBSETTING IF STATEMENT

Subsetting IF statements are used to exclude some observations from the data set. Only observations meeting the IF criteria are included in the SAS data set. Some examples are

```
if carsize = 'small';
if 25 < mpg < 35;
if country = 'japan' or country = 'us';
if country ^= 'germany';
if (country = 'us') and (carsize = 'large');
```

The usual arithmetic operators work in SAS: add (+), subtract (−), multiply (*), divide (/), and exponentiate (**). For ≠, use ^=, ¬=, ˜=, or NE depending on the type of keyboard you have. You can use parentheses to group operations. Parentheses are evaluated first, then exponentiation, then multiplication and division, and then addition and subtraction. For example, $5+3*4**2 = 5+3*16 = 5+48 = 53$ and $(5+3)*4**2 = 8*16 = 128$.

IF...THEN...ELSE STATEMENTS

IF...THEN...ELSE statements are used to create new variables where the value of the new variable is based on the value of an existing variable. This statement can also be used to change the value of a variable. IF...THEN is used in one SAS statement. ELSE is used in subsequent statements to make SAS work more efficiently. Keep in mind that the original data set is not changed; only the SAS data set reflects the changes made.

In the following set of code, an ELSE statement was not used. SAS looks at each statement to see if the IF condition holds. Assume that the first observation had a mpg value of 33. SAS would evaluate the first IF statement and find that it was not true. It would then go to the next IF statement and find it was true, giving the value 'good ' to the variable mileage. It would then go to the next IF statement and find that it was not true, and so on.

```
if 50 <= mpg       then mileage = 'great';
if 30 <= mpg < 50 then mileage = 'good ';
if 20 <= mpg < 30 then mileage = 'okay ';
if       mpg < 20 then mileage = 'bad  ';
```

A better, more efficient way to write this code is to use ELSE statements. SAS keeps evaluating the statement until it finds one that is true. It then skips the rest of the statements. In the following code, it would evaluate the first statement and find it was not true (for mph = 33). It would then go to the second IF statement, and since it is true, it would skip the remaining two statements. Notice how spacing makes the code easier to read.

```
if       50 <= mpg       then mileage = 'great';
else if 30 <= mpg < 50 then mileage = 'good ';
else if 20 <= mpg < 30 then mileage = 'okay ';
else if       mpg < 20 then mileage = 'bad  ';
```

Another way to write the code uses the logic of IF...THEN...ELSE statements.

```
if       50 <= mpg then mileage = 'great';
else if 30 <= mpg then mileage = 'good ';
else if 20 <= mpg then mileage = 'okay ';
else if       mpg then mileage = 'bad  ';
```

When creating character variables this way, you need to be careful. SAS sets the length of the variable the first time it is evaluated. That means that the first IF statement would set the length of the variable MILEAGE. All the values of MILEAGE have a length of 5, which is the length of the longest value 'great.' If spaces were not included in the shorter values, the length of MILEAGE could be 3, for the shortest value 'bad.' Then the possible values of MILEAGE would be gre, goo, oka, and bad.

USING DATA FROM OUTSIDE FILES

Many data sets exist in a computer file somewhere. If you collect a set of data that has many observations, you may want to put it in a separate file that you can easily access. It would be a waste of time and a duplication of effort if you had to retype the data in your SAS program. Fortunately, you don't have to do this. You can just tell SAS where your data set actually resides.

When the data is not in the SAS program and read by a CARDS statement, it exists in an external file. This means the file is external to the SAS program. The file can reside on a main frame computer, on a hard disk, or on a floppy disk. Wherever it is, it is referred to as an external file.

Two statements work together to allow you to read your data from this external file: the FILENAME statement and the INFILE statement.

The FILENAME Statement

This statement is used to link an outside file to the SAS program by giving it a SAS name. This name is like an alias: SAS will always use the alias whenever it needs to refer to the file. The FILENAME statement must precede the DATA step that reads the file. In the following statement, the file 'car.dat' is associated with the SAS name DATAIN. How you refer to these external data sets is site specific. You will need to check with a computer systems person to learn how to refer to these files at your location.

```
filename datain 'car.dat';
```

The INFILE Statement

The INFILE statement is used in the DATA step before the INPUT statement. It lets SAS know that the data will be read from an external file instead of being input with a CARDS statement. Here is an example.

```
data cars;
   infile datain;
   input mpg;
```

PROC SORT

To rearrange the order of the SAS data set, you use PROC SORT. How the sort is executed depends on a BY statement. If you want the mileage data sorted by mpg, you would use

```
proc sort data = cars;
   by mpg;
```

It is also possible to sort by more than one variable. The first variable listed after the BY statement in PROC SORT does the initial sort. The next variable sorts within the first variable; the third variable listed sorts within the first two; and so on. For instance, you may want to see the car listed in mpg order for each country. The following statement will list the data by country in alphabetical order. Within each country, the data will be from small mpg to largest.

```
proc sort data = cars; by country mpg;
```

You will get different results if you sort BY MPG COUNTRY. After this sort, the data will be grouped from smallest to largest mpg. For each mpg value, the data will be in alphabetical order by country.

The sort order for character data depends on the collating sequence at your computer site. One sequence sorts blanks first, then numbers, then capital letters, then small letters. The other sequence sorts blanks first, then small letters, then capital letters, and then numbers. Special characters, like @, !, and &, are handled differently by each sorting sequence. You should refer to a SAS manual if you need to sort special characters.

PROC SORT General Form

```
proc sort data = data set;
   by variables <options>;
```

BY Statement Option	Description
descending	The default is for PROC SORT to sort from smallest to largest (ascending). To sort a variable from largest to smallest use DESCENDING before the variable name.

EXAMPLE 2.1

Dr. Redding has data from a skin study where some patients received a new drug and some received a placebo. Patients were from different clinics. Skin thickening, skin mobility, and patient assessment were measured at the beginning of the study and again at the end of the study. (See the Appendix for a complete description of this data.) The data is in a file called 'sclero.dat' and has 76 observations.

He wants to look at the data for clinics 45 through 50, where the patients showed some improvement on the patient assessment variable. (The higher the number, the worse the patient is.) He wants the information to be grouped by whether the patient took a placebo or the new drug. He also wants to compute this improvement.

See Figure 2.1 for the output.

SAS Program

```
filename datain 'sclero.dat';

data one;
   infile datain;
   input clinic id drug thick1 thick2 mobilty1
         mobilty2 assess1 assess2;
   improve = assess1 - assess2;
   if improve > 0;
   if 45 <= clinic <= 50;
run;

proc sort;
   by drug;
run;

proc print;
   by drug;
run;
```

New Statements

```
filename datain 'sclero.dat';
```
This statement takes the external file 'sclero.dat' and associates it with the SAS name DATAIN.

```
infile datain;
```
This statement *must* come before the INPUT statement. It means that the data is coming from the external file associated with DATAIN in the FILENAME statement.

```
improve = assess1 - assess2;
```
IMPROVE is the name of a new variable that represents the difference between the first assessment score and the second assessment score.

```
if improve > 0;
if 45 <= clinic <= 50;
```
These are subsetting IF statements. Another way to subset for the clinic variable would be the statement

```
              if clinic = 48 or clinic = 49 or clinic = 50;
```

```
proc sort;
by drug;
```
When sorting the data set, you must tell SAS which variables to sort by.

```
proc print;
```

by drug;   The SAS output will contain a listing of the data grouped by drug. *You cannot use a BY statement with any procedure without first having sorted the SAS data set by that variable.*

Discussion of Output

Notice in the output for Example 2.1 that the data set is grouped by the different values of the variable drug, with drug=1 first. Only observations with clinic numbers between 45 and 50 are part of the data set. The dots (.) under some of the variables refer to missing data. There were no measurements on those variables for some people.

Figure 2.1   SAS List File for Example 2.1

```
                           The SAS System13:02 Wednesday, April 28, 1993

-------------------------------- DRUG=1 --------------------------------

 OB  CLINIC  ID  THICK1  THICK2  MOBILTY1  MOBILTY2  ASSESS1  ASSESS2   IMPROVE

  1    45     4    38       .       184       195        7        6         1
  2    45     8    26       .       255       211        5        2         3
  3    46     3     7       7       440       444        4        3         1
  4    46     8    17      18       191       192        7        5         2
  5    47     2    28      28       191        .         9        8         1
  6    48    24    26      30       211       310        9        7         2
  7    49     7     6       6       531       521        2        1         1
  8    50     8    19       9       432       398        5        1         4


-------------------------------- DRUG=2 --------------------------------

 OB  CLINIC  ID  THICK1  THICK2  MOBILTY1  MOBILTY2  ASSESS1  ASSESS2   IMPROVE

  9    45     6    24       .       472       537        2        1         1
 10    46     2     9       9       347       309        5        2         3
 11    46    14    12      15       313       373        9        5         4
 12    46     7     9      11       359       395        5        3         2
 13    47     1     7       7       383        .         5        2         3
 14    48     2    18       7       444       563        6        4         2
 15    49     4    18      18       378       373        3        2         1
 16    49     3     7       7       675       636        3        1         2
 17    50    10    17       9       510       448        3        1         2
```

SET STATEMENT

You can have many SAS data sets in one program. It is possible to use a SAS data set as the basis for creating a new SAS data set. The SET statement is used in a DATA step to refer to another SAS data set in the program. That original data set still exists and can be used by SAS. The observations in the original data set are evaluated by the new data set using the SET statement.

EXAMPLE 2.2

In a continuation of Example 2.1 on skin measurements, suppose Dr. Redding wants to do a different analysis for the observations from clinic 49. He could create a new SAS data set that contains only the observations for that clinic by using a subsetting IF statement and a SET statement.

SAS Program

```
data clinic49;
   set one;
   if clinic = 49;

proc print data = clinic49;
run;
```

New Statements

set one;   The data are read in from the SAS data set ONE instead of from a CARDS statement or from an outside file. The original data set ONE still exists. Since two data sets are now available for PROCs, it is necessary to specify which data set should be used. The default is that SAS will use the most recently created data set. It is always best to specify which data set to analyze.

Discussion of Output

The output (see Figure 2.2) reflects which observations are in the SAS data set CLINIC49. There are three observations.

Figure 2.2   SAS List File for Example 2.2

```
                            The SAS System13:10 Wednesday, April 28, 1993


OB  CLINIC  ID  THICK1  THICK2  MOBILTY1  MOBILTY2  ASSESS1  ASSESS2   IMPROVE

 1    49     7     6       6       531       521        2        1         1
 2    49     4    18      18       378       373        3        2         1
 3    49     3     7       7       675       636        3        1         2
```

PROBLEMS

The files for these problems are described in detail in the Appendix.

2.1   The file utility.dat contains a monthly record of telephone, electricity, and fuel costs for several years.

a)   Create a new variable for the total expenses per month. Print the data set.
b)   Print telephone costs by year and in month order. (HINT: Sorting the current month variable will put months in alphabetic order. Sorting on a numeric variable for month will put months in the correct order.)
c)   Print telephone costs by month and in year order. (Same HINT as for (b).)
d)   Print this data set for 1992 only.
e)   Print this data set for the months of January, February, and March by year.
f)   How are (b) and (c) different? When would you want to look at data this way?

2.2   The file china#1.dat contains export and import information (in dollars) by year.

a)   Create a new variable that reflects the trade balance: exports − imports. Print the data set.
b)   Print this data set by decade.
c)   Print this data set for year 1980 and later in year order.
d)   Comment on the data. What is happening to exports, imports, and the trade deficit?

2.3   The file well#1.dat contains nitrate, zinc, and TDS (total dissolved solids) levels from a well being monitored for contamination.

a)   The units are in milligrams per liter (mg/l). Change the units to grams per liter for TDS. Print the variables month, day, year, TDS in mg, and TDS in grams.
b)   Print TDS, nitrate, and zinc in that order for 1991.

2.4   The file handinj.dat contains the costs (in Irish pounds) and lost work days due to hand injuries for workers in Dublin, Ireland.

a)   Change Irish pounds to U.S. dollars (1 Irish pound = 1.54 U.S. dollars).
b)   Print the data set by days lost in decreasing cost (dollars) order.
c)   Print the data set for work injuries in order of days lost.
d)   What are the ID numbers of individuals with the most costly claims? Who lost the most days of work?

YOU SHOULD NOW KNOW

how to read from an outside file

how to create new variables in the DATA step

how to use a subset of the data

how to sort the data

how to use the SET statement

Data Management
MODULES NEEDED 1, 2

For small data sets and straightforward analyses, it is simple to create a SAS data set and apply one of the PROCs. However, there are times when there is a lot of data, the data is not in a good form for analysis, or the analysis you have in mind seems rather complicated in a data handling sense.

SAS has many useful techniques for dealing with data and data sets. It is capable of doing almost anything you would like to do to your data. SAS creates special variables that can be useful. It has several ways of combining data sets. Just as you can read data from a file, you can also create a new file.

In this module, you will learn about some of the data management tools available in SAS. If you find these very useful, you may want to consult a SAS manual to understand the full data management tools SAS makes available. The information in this module is only the tip of the iceberg.

HOW SAS "THINKS"

SAS takes your raw data and transforms it into a SAS data set that can be analyzed using SAS PROCs. How does it do this?

For each DATA step, SAS creates a data vector. The vector includes all the variables mentioned in the DATA step, whether they are on the INPUT statement or on other SAS statements that create variables. The vector also includes the special SAS variable _N_. SAS sets up this vector before reading in any data. The vector represents one observation containing all the variables mentioned in the DATA step.

Initially, all the values of the variables are set to missing, except for the variable _N_, which has the value 1. _N_ keeps track of how many times SAS evaluates the statements in the DATA step. SAS then reads the data values for the first observation from the INPUT statement and computes any new variables from other statements. It evaluates statements that subset the data in some way, like the IF statement. Then, SAS writes that observation to the SAS data set. It then goes back to the beginning, sets all the values to missing except for _N_, which now is 2, reads values for the second observation, writes it to the SAS data set, and so on.

This process continues until all the observations have been written to the SAS data set.

LAG FUNCTION

SAS handles data one observation at a time. Sometimes it is useful to use a value from the previous observation. The LAG function takes a value from a variable in the previous observation and assigns it to a new variable in the current observation.

The form of this function is

```
lag(variable) = newvariable
```

Assume you have a data set containing daily high temperatures. You want to compute the change in temperature. The data contains the variables day and temperature.

```
Jan  1      33
Jan  2      35
Jan  3      45
```

The change in temperature from Jan 1 (the first observation) to Jan 2 (the second observation) is +2, and the change from Jan 2 to Jan 3 (the third observation) is +10. To do the arithmetic for the change on Jan 2, both temperature values (33 and 35) must be available to the DATA step. The following statement adds a new variable LAG_TEMP to the variables DAY and TEMP:

```
lag_temp = lag(temp);
```

The DATA step now consists of day, temperature, and lag temperature. Notice that the first value of LAG(TEMP) is missing: There is no previous value to lag.

```
Jan  1      33      .
Jan  2      35      33
Jan  3      45      35
```

To compute the change in temperature, use the following statement:

```
change = temp - lag_temp;
```

DROP AND KEEP STATEMENTS

These two statements can be used to exclude certain variables from the SAS data set or to make sure only certain variables are included in the SAS data set. Both of them cannot be used in the same DATA step. If you plan to drop more variables than you are keeping, use the KEEP statement. If you plan to keep more variables than you drop, use the DROP statement. They have the form

```
DROP variable list;
KEEP variable list;
```

Why wouldn't you keep all variables in the data set? It takes computer memory resources to keep track of a lot of variables. If some of them are not going to be used for any analysis, you may not need them in the data set. Perhaps some variables are used only to create new variables that will be analyzed. Then, keep only those variables that are necessary for the analysis. This saves computer resources and can make programming easier.

Some PROCs will do an analysis on all variables in a data set if you do not specify which ones to use. For instance, PROC PRINT will print all variables in the data set if you do not include a VAR statement. If the SAS data set has only the variables you really need, you would not have to use a VAR statement with PROC PRINT.

COMBINING DATA SETS

There are several ways to combine data sets in SAS. Those described here are concatenating, interleaving, one-to-one merging, and match merging. Although it is possible to combine two or more data sets, this module will focus on combining two data sets only. Since all these ways result in a different final data set, the examples will utilize the same two data sets, YR1992 and YR1993, which contain temperature and fuel cost information for a few days in January.

YR1992				YR1993	
Day	Temp	Weather		Day	Temp
Jan 1	33	sun		Jan 1	25
Jan 2	35	sun		Jan 2	20
Jan 3	45	clouds		Jan 3	35
Jan 5	28	snow		Jan 6	40
				Jan 8	39

Concatenating Data Sets

To concatenate data sets, you use a SET statement. The total number of observations in the final data set equals the sum of the observations in the data sets that are being combined. The observations in the first data set listed are read first, then the observations from the second data set, and so on. The number of variables in the combined data set will be equal to the number of different variables in the data sets to be combined. Remember that the variable TEMP is not the same as a variable named TEMPS.

Assume the variables in data set YR1992 are DAY, TEMP, and WEATHER, and the variables in YR1993 are DAY and TEMP. The statement to concatenate them is

```
data combined;
   set yr1992 yr993;
```

Data set COMBINED has nine observations and missing values for WEATHER for the data from YR1993.

```
                   COMBINED
         Day      Temp     Weather
      Jan  1       33      sun
      Jan  2       35      sun
      Jan  3       45      clouds
      Jan  5       28      snow
      Jan  1       25
      Jan  2       20
      Jan  3       35
      Jan  6       40
      Jan  8       39
```

Interleaving Data Sets

Interleaved data sets have the same number of observations as concatenated data sets. However, the order is different. A BY statement is used to determine what variable common to both data sets is used for the interleaving process. Remember that to use a BY statement, each data set must first be sorted by that variable.

Assume that data set YR1992 has variables DAY, TEMP, and WEATHER, and YR1993 has variables DAY and TEMP. The statement to interleave them by DAY is

```
data new;
   set yr1992 yr1993;
   by day;
```

Data set NEW has nine observations and missing values for WEATHER for the data from YR1993. Notice how the order differs from concatenation.

```
                   NEW
         Day      Temp     Weather
      Jan  1       33      sun
      Jan  1       25
      Jan  2       35      sun
      Jan  2       20
      Jan  3       45      clouds
      Jan  3       35
      Jan  5       28      snow
      Jan  6       40
      Jan  8       39
```

One-to-One Merging of Data Sets

To do a one-to-one merge, you use a MERGE statement without a BY statement. The number of observations in the resulting data set is the largest number in the data sets that are merged. The number of variables is equal to the total number of variables in those data sets. If each of the data sets has a variable with exactly the same name, the value of the variable will be that of the data set listed last in the MERGE statement.

Assume that data set YR1992 has variables DAY, TEMP, and WEATHER, and that YR1993 has variables DATE (previously named DAY) and TEMP. The statements for one-to-one merging are

```
data together;
   merge yr1992 yr1993;
```

Data set TOGETHER has five observations, four variables, and missing values for DAY and WEATHER for the last observation. Notice that the value of TEMP came from YR1993 because it was listed last on the MERGE statement.

		TOGETHER		
Day	Temp	Weather	Date	
Jan 1	25	sun	Jan 1	
Jan 2	20	sun	Jan 2	
Jan 3	35	clouds	Jan 3	
Jan 5	40	snow	Jan 6	
	39		Jan 8	

Match Merging of Data Sets

Match merging combines two or more data sets with the use of a MERGE statement and a BY statement. Observations in sorted data sets are matched together through the value of a common variable. The resulting data set will have the number of observations equal to the sum of the largest number of observations in each BY group.

Assume that data set YR1992 has variables DAY, TEMP92 (previously called TEMP), and WEATHER, and that YR1993 has variables DAY and TEMP93 (previously called TEMP). The statements for match merging are

```
data merged;
   merge yr1992 yr1993;
   by day;
```

Data set MERGED has five observations, four variables, and missing values for TEMP92 and WEATHER for the last observation since there was no Jan 8 in YR1992.

	MERGED		
Day	Temp92	Weather	Temp93
Jan 1	33	sun	25
Jan 2	35	sun	20
Jan 3	45	clouds	35
Jan 5	28	snow	
Jan 8			39

In another example of match merging, let DATA1 and DATA2 be as shown and merged by COLOR according to the SAS code. Notice the resulting DATA3.

DATA1	
blue	10
blue	23
blue	15
red	11
red	19
red	24

DATA2	
blue	$14.00
red	$20.00

```
data data3;
   merge data1 data2;
   by color;
```

DATA3		
blue	10	$14.00
blue	23	$14.00
blue	15	$14.00
red	11	$20.00
red	19	$20.00
red	24	$20.00

COMMENT LINES

It is easy to include comments in SAS programs. Including comments is helpful when programs are long or when you want to remember what a particular group of code does. There are two ways to include comments. One is to begin a SAS statement with an asterisk (*) and end the statement in the usual way with a semicolon. There other way is to begin the comment with /* and end with */. Here are examples of comments.

```
data sales;
   input item cost;
 * compute sales tax;
   tax = cost*0.0625;

data sales;
```

```
input item cost;
tax = cost*0.0625   /* compute sales tax */ ;
```

LIMITING THE NUMBER OF OBSERVATIONS READ FROM EXTERNAL FILES

Some data files contain header information that should be skipped. When SAS encounters information that does not fit the description in the INPUT statement, it returns error messages that say the data is invalid. It is easy to skip these extra lines with a FIRSTOBS statement. FIRSTOBS points to the first observation that SAS should read. It is used on the INFILE statement. For example, the following lines tell SAS to start reading the fifth observation of the data file.

```
data money;
   infile banks firstobs=5;
   input bank $ balance type $;
```

Sometimes you may want to limit the number of observations read from a data file. This is a particularly nice technique when you are writing a program that uses a large data set. It is always good to make sure that you get what you wanted to get and not what you told SAS to do. To check, you can use a smaller number of observations, print out the results, and see whether the program is working the way you intended. To do this, you use an OBS statement on the INFILE statement, just like the FIRSTOBS command. It is possible to use both of them together. The following code sets the number of observations to 25.

```
data money;
   infile banks obs=25;
   input bank $ balance type $;
```

The following code tells SAS to begin with the fifth observation of the data file and have only 25 observations.

```
data money;
   infile banks firstobs=5 obs=25;
   input bank $ balance type $;
```

SETTING THE SIZE OF THE OUTPUT

Each SAS system is set up to print the output on a particular size of paper. To override the system defaults, you can use the LINESIZE and PAGESIZE options. These appear on an OPTIONS statement that would be the first statement in the SAS program. Setting the size of the page will result in fewer blocks of output being split across pages, which is difficult to read.

LINESIZE sets the number of characters that will print across the width of the page. Acceptable values range from 64 through 256.

PAGESIZE sets the number of lines that will print down the length of the page. Acceptable values range from 15 through 32767.

The following line sets LINESIZE and PAGESIZE values, which usually work well for an 8½-by-11 inch sheet of paper.

```
options linesize=64 pagesize=55;
```

THE PUT STATEMENT

It is possible to write information on the SAS LOG file and to create new data files using the PUT statement. It is the complement of the INPUT statement and uses many of the same formatting commands. PUT statements appear in DATA steps.

To write information to the SAS LOG, use PUT followed by variable names or text. Here are some examples.

```
data one;
   input name $ age;

put name age;
put @5 'the name is ' name ' and the age is ' age;
put 'This program is working.'
```

To write information to an external file, you need a FILE statement and a PUT statement. This is useful if you would like to save the computations or changes you made in the SAS program to a data file. The FILE statement is the complement of the INFILE statement: It must be defined first with a FILENAME statement. If a FILE statement is not used, SAS writes to the LOG file.

```
filename dataout 'newfile.dat';

data one;
   input name $ age;
file dataout;
put @5 name @20 age;
```

PROBLEMS

The files for these problems are described in detail in the Appendix.

3.1   Include a comment in your code when using file china#1.dat to print a data set containing

a)   the change in exports by year. Let change = this year − last year.

b)   the change in imports by year. Let change = this year − last year.

3.2   Include a comment in your code when using the file utility.dat to print a data set containing

a)   the monthly change in telephone costs for 1990.

b)   the yearly change in fuel costs for January.

3.3   Create a data set containing three variables and ten observations. Create a second data set also containing three variables and ten observations where one of the variables is the same as in the first data set. Merge the two data sets together by

a)   concatenating.

b)   interleaving by a common variable.

c)   match-merging by a common variable.

YOU SHOULD NOW KNOW

how to combine data sets

how to create LAG variables

how to include comments in SAS programs

how to set the page size for output

SAS Functions

MODULES NEEDED 1, 2

SAS has many functions for working with data values. Some perform arithmetic and trigonometric transformations. Some work with character variables. Others compute probabilities from different distributions. This module describes some of the many functions available in SAS.

NUMERIC FUNCTIONS

Here is a list and description of some of the numeric functions.

Function	Description
abs(*argument*)	Returns the absolute value of the argument.
exp(*argument*)	Raises the number e to the power of the argument.
int(*argument*)	Returns the integer portion of the argument.
log(*argument*)	Computes natural log of argument (argument must be > 0).
log10(*argument*)	Returns the base 10 log of argument (argument must be > 0).
round(*argument*, *round-off-unit*)	Rounds the argument to the nearest value of the round-off-unit. For example, the statements x=33.2217; y=round(x,.001); would result in y equal to 33.222.
sqrt(*argument*)	Returns the positive square root of the argument.

arcos(*argument*)	Computes arccosine. Argument must be between −1 and +1.
arsin(*argument*)	Computes arcsine. Argument must be between −1 and +1.
atan(*argument*)	Computes arctangent.
cos(*argument*)	Computes cosine of argument, which must be in radians.
cosh(*argument*)	Computes the hyperbolic cosine.
sin(*argument*)	Computes the sine of argument, which must be in radians.
sinh(*argument*)	Computes the hyperbolic sine.
tan(*argument*)	Computes the tangent of argument, which must be in radians.
tanh(*argument*)	Computes the hyperbolic tangent.

CHARACTER FUNCTIONS

Here is a list and description of some of the character functions.

Function	Description
left(*argument*)	Left-aligns argument.
right(*argument*)	Right-aligns argument.
substr(*argument, position, n*)	Extracts a substring of argument beginning with the character at specified position and having length n.
trim(*argument*)	Removes trailing blanks from argument.

PROBABILITY FUNCTIONS

Here is a list and description of some of the probability functions.

Function	Description
poisson(*m,n*)	Returns $P(X \leq n)$, where X is a Poisson random variable and $m \geq 0$ is the mean.
probbnml(*p,n,m*)	Returns $P(X \leq m)$, where X is a binomial random variable, $0 \leq p \leq 1$, and $n > 0$ is the number of trials.
probchi(*x,df*)	Returns $P(X \leq x)$, where X is a chi-square random variable, df is the degrees of freedom, and $x \geq 0$.
probf(*x,ndf,ddf*)	Returns $P(X \leq x)$, where X is an F random variable, ndf is the numerator degrees of freedom, ddf is the denominator degrees of freedom, and $x \geq 0$.
probhypr(*N,K,n,x*)	Returns $P(X \leq x)$, where X is a hypergeometric random variable $N > 1$ is the population size, $0 \leq K \leq N$ is the number in the population with a special characteristic, $0 \leq n \leq N$ is the sample size, and $\max(0, K+n-N) \leq x \leq \min(K, n)$.
probnorm(*x*)	Returns $P(X \leq x)$, where X is a normal random variable with $\mu = 0$ and $\sigma = 1$.
probt(*x,df*)	Returns $P(X \leq x)$, where X is a t random variable with df degrees of freedom.

EXAMPLE 4.1

Sandra Smith would like to verify some binomial probabilities. In particular, she would like to duplicate part of the table in her statistics book. She chooses n=5 and p=0.4.

SAS Program

```
data one;
   input x;
   n = 5;
   p = 0.4;
   cdf = probbnml(p,n,x);
cards;
0
1
2
3
4
5
;
run;

proc print data = one;
   var x cdf;
run;
```

Discussion of Output

You can easily verify that the PROC PRINT table in Figure 4.1 is the cumulative distribution function for a binomial random variable with n=5 and p=0.4.

Figure 4.1   SAS Output for Example 4.1

```
The SAS System                       1
                         11:43 Tuesday, April 12, 1994


              OBS    X      CDF

               1     0    0.07776
               2     1    0.33696
               3     2    0.68256
               4     3    0.91296
               5     4    0.98976
               6     5    1.00000
```

PROBLEMS

The files for these problems are described in detail in the Appendix.

4.1   Use the file well#1.dat and use SUBSTR to create new variables for month and day by reading in the month/day field as one variable. Print the data.

4.2   Create a data set that has one variable with values 2.7, −6.9, 3.4, −0.5, and 1.3. Create a new variable that is the absolute value of these data and then take the square root. Can you do this in one step by embedding the commands?

4.3   Let X be a binomial random variable with n=13 and p=0.23. Create a data set of the possible values of X. Print a data set containing

     a) $P(X \le x)$.
     b) $P(X > x)$.
     c) $P(X < x)$.

4.4   Create a data set of the possible values for a binomial random variable X, letting n=5 and p=0.40.

     a)   Compute the cumulative distribution function, $P(X \le x)$.
     b)   Compute the mass function, $P(X=x)$.
     c)   Verify these results by checking the binomial table in your statistics book.

4.5   Let X be a normal random variable with $\mu=12.6$ and $\sigma=2.3$.

     a)   Compute $P(X < 10)$.
     b)   Compute $P(7.6 < X < 15)$.

4.6   Choose ten entries in the normal table in your statistics book and verify that they are correct.

YOU SHOULD NOW KNOW

how to use numeric, character, and probability functions

Descriptive Statistics I
MODULES NEEDED   1, 2

We use statistics to help describe data. In particular, we compute means, standard deviations, and percentiles, and we draw histograms and stemplots. Several SAS procedures compute descriptive statistics.

Computer output can be difficult to read, particularly because variable names can be only eight characters long. Sometimes it is difficult to assign an eight-character variable name that is meaningful. SAS provides a way to make the output results easier to read by using LABEL and TITLE statements.

In this module, you will learn how to use PROC UNIVARIATE to calculate descriptive statistics and how to make your output easier to read.

LABEL STATEMENTS

The LABEL statement assigns labels to variables. These labels act like aliases and print alongside the variable name on the output. Labels can contain up to 40 characters. Any characters are valid for the text inside the single quotes. If the label contains a single quote or apostrophe, use two single quotes at that place. It is possible to create labels for more than one variable at a time. LABEL statements are valid in DATA steps and PROC steps. If found in a PROC step, the label is valid only for that procedure. If found in a DATA step, the label is valid throughout the program unless it is changed by a subsequent LABEL statement. Here is an example.

```
label time     = 'time needed to complete exam'
      teacher  = 'teacher''s name'
      score    = 'exam score'
      ;
```

TITLE STATEMENTS

TITLE statements appear at the top of each page of output. You can create more than one title on a page by numbering the TITLE statements TITLE1, TITLE2, and so on. TITLE statements are in effect until changed in subsequent TITLE statements. Changing TITLE number n deletes titles with a number higher than n. If, for example, a program uses

TITLE1, TITLE2, and TITLE3 statements, and TITLE2 is changed, TITLE1 is still in effect, but TITLE3 is lost. Titles are centered by default.

PROC UNIVARIATE

PROC UNIVARIATE generates descriptive statistics: mean, standard deviation, percentiles or quantiles (1%, 5%, 10%, 25%, 50%, 75%, 90%, 95%, 99%), minimum value, and maximum value. It also prints the five largest and five smallest observations in the data set.

The PLOT option will generate a stem-and-leaf plot, a box plot, and a normal probability plot. For large data sets, a horizontal bar chart may be produced instead of a stem-and-leaf plot.

In the box plot, the top and bottom of the box are the 75th and 25th percentiles, respectively. The median is denoted by a bar, and the mean is denoted by a plus sign. Lines extend from the top and bottom of the box to a distance of at most 1½ times the interquartile range (IQR). Observations (possible outliers) greater than 1½ IQR and less than 3 IQR are represented by a circle; those greater than 3 IQR are represented by an asterisk.

In the normal probability plot, data points are represented by asterisks, and plus signs denote a straight line for reference. If the data are normally distributed, the data points should fall along the straight line.

The NORMAL option generates a statistic to test for normality and its p-value.

The ID statement specifies a variable whose value is printed next to the smallest and largest observations. This could be an item number, a name, a survey number, or other variable that would make it easier to investigate extreme observations.

PROC UNIVARIATE calculates many other statistics that are not discussed in this manual.

PROC UNIVARIATE General Form

```
proc univariate data=data set <options>;
   by variables;
   var variables;
   id variables;
```

UNIVARIATE Option	Description
normal	Computes a statistic to test whether the data comes from a normal distribution. If $n \leq 2000$, the Shapiro-Wilk W is computed. If $n > 2000$, the Kolmogorov D is computed.
plot	Produces a stem-and-leaf plot (or a horizontal bar chart if n is large), a box plot, and a normal probability plot.

EXAMPLE 5.1

Continuing with Example 2.1 about data from a skin study, Dr. Redding wanted to generate some descriptive statistics. In particular, he wanted

- to have a longer description for each variable
- to describe the shape of the distribution for assessment improvement for the group that took the drug and the group that took the placebo
- to compute the sample means and standard deviations for these two groups.
    See Figure 5.1 for part of the output.

SAS Program

```
filename datain 'sclero.dat';

data one;
   infile datain;
   input clinic 1-2 id 4-5 drug 8 thick1 11-12 thick2 15-16
      mobilty1 19-21 mobilty2 24-26 assess1 29 assess2 32;
   improve = assess1 - assess2;

   label
      id       = 'Patient ID Number'
      drug     = 'Drug (1) or Placebo (2)'
      thick1   = 'Skin Thickening at 1st Visit'
      thick2   = 'Skin Thickening at 2nd Visit'
      mobilty1= 'Skin Mobility at 1st Visit'
      mobilty2= 'Skin Mobility at 2nd Visit'
      assess1  = 'Patient Assessment at 1st Visit'
      assess2  = 'Patient Assessment at 2nd Visit'
      improve  = 'Improvement in Assessment Score'
   ;
run;

proc sort;
   by drug;
run;

proc univariate plot;
   by drug;
   var improve;
   id id;
title1 'Scleroderma Study';
title2 'Conducted Fall 1989';
run;
```

New Statements

```
label id= 'Patient ID Number'
```

```
drug    = 'Drug=1    Placebo=2'
thick1  = 'Skin Thickening at 1st Visit'
thick2  = 'Skin Thickening at 2nd Visit'
mobilty1= 'Skin Mobility at 1st Visit'
mobilty2= 'Skin Mobility at 2nd Visit'
assess1 = 'Patient Assessment at 1st Visit'
assess2 = 'Patient Assessment at 2nd Visit'
improve = 'Improvement in Assessment Score'
;
```
                        This single LABEL statement assigns labels to nine variables.

```
title1 'Scleroderma Study';
title2 'Conducted Fall 1989';
```
                        TITLE1 will appear above TITLE2 on all pages of output.

```
proc univariate plot;
var improve;
id id;
```
                        The value of the variable ID, which is patient ID, will print next to
                        the smallest and largest observations. Notice that SAS recognizes the
                        first ID as a keyword and the second ID as a variable name.

Discussion of Output

Figure 5.1 shows the output for IMPROVE at one drug value. SAS produced two pages of output.

On the first page, the number of observations (N)=31, the sample average (Mean)=0.645, the sample standard deviation (Std Dev)=2.026, the total of the observations (Sum)=20, and the sample variance (Variance)=4.103.

Several quantiles are computed: first quartile (Q1)=−4, median (Med)=0, and third quartile (Q3)=2. The range=9, the interquartile range (Q3−Q1)=2, and the mode=0.

Next on the output is a list of the lowest values and highest values in the data set. Patient 42 had the lowest value of −4, and patient 15 had the highest improvement value of 5.

The next three lines say that there was one missing value and that 3.13% of the data is missing.

On the second page of output, the stem-and-leaf plot shows the shape of the improvement values had a peak at 0 and is not symmetrical: There are more positive observations than negative ones, indicating that most people did not improve completely. On the normal probability plot, some of the data points (the asterisks) do not fall in a straight line (on the pluses), indicating that the data may be nonnormal.

Figure 5.1   SAS Output for Example 5.1 for Variable IMPROVE and DRUG = 1

```
                        Scleroderma Study   1
                         Conducted Fall 1989
                    13:02 Wednesday, April 28, 1993


--------------------- Drug (1) or Placebo (2)=1 ---------------------

                       Univariate Procedure

Variable=IMPROVE       Improvement in Assessment Score

                              Moments

        N               31  Sum Wgts         31
        Mean      0.645161  Sum              20
        Std Dev   2.025642  Variance   4.103226
        Skewness  0.186157  Kurtosis   0.358925
        USS            136  CSS        123.0968
        CV        313.9745  Std Mean   0.363816
        T:Mean=0  1.773317  Prob>|T|     0.0863
        Num ^= 0        19  Num > 0          13
        M(Sign)        3.5  Prob>|M|     0.1671
        Sgn Rank        42  Prob>|S|     0.0919


                          Quantiles(Def=5)

        100% Max         5     99%          5
         75% Q3          2     95%          4
         50% Med         0     90%          4
         25% Q1          0     10%         -1
          0% Min        -4      5%         -3
                                1%         -4

        Range            9
        Q3-Q1            2
        Mode             0


                             Extremes

        Lowest    ID      Highest    ID
           -4(    42)        3(      36)
           -3(    26)        4(       5)
           -2(    50)        4(      29)
           -1(    73)        4(      75)
           -1(    30)        5(      15)


                  Missing Value            .
                  Count                    1
                  % Count/Nobs          3.13
```

Figure 5.1 cont.   SAS Output for Example 5.1

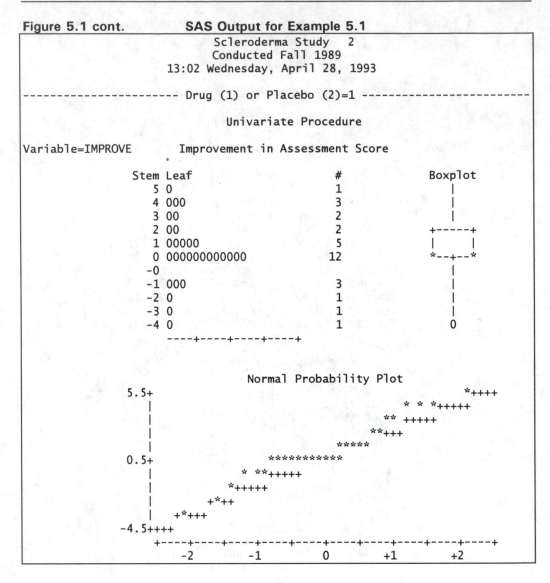

```
                    Scleroderma Study    2
                     Conducted Fall 1989
                 13:02 Wednesday, April 28, 1993

---------------------- Drug (1) or Placebo (2)=1 ---------------------

                      Univariate Procedure

Variable=IMPROVE      Improvement in Assessment Score

         Stem Leaf                        #          Boxplot
            5 0                            1             |
            4 000                          3             |
            3 00                           2             |
            2 00                           2          +-----+
            1 00000                        5          |     |
            0 000000000000               12          *--+--*
           -0                                           |
           -1 000                          3             |
           -2 0                            1             |
           -3 0                            1             |
           -4 0                            1             0
              ----+----+----+----+
```

```
                    Normal Probability Plot
        5.5+                                              *++++
           |                                      * * *+++++
           |                                   ** +++++
           |                                 **+++
           |                              *****
        0.5+                    ***********
           |              * **+++++
           |             *+++++
           |          +*++
           |       +*+++
       -4.5++++
          +----+----+----+----+----+----+----+----+----+----+
              -2        -1         0        +1        +2
```

EXAMPLE 5.2
Look at the following code using the scleroderma data of Example 5.1 and how the title statements are used.

SAS Code

```
proc univariate;
   var improve;
title1 'Scleroderma Study';
```

```
title2 'Improvement in Patient Assessment';
run;

proc univariate;
   var thick1;
title2 'Patient Skin Thickness at 1st Visit';
run;

proc univariate;
   var thick2;
title2 ;
run;
```

Discussion

For the first procedure, TITLE1 will be centered above TITLE2. For the second procedure Scleroderma Study will be centered above the new TITLE2. For the last procedure, only TITLE1 will appear since TITLE2 is blank.

PROBLEMS

The files for these problems are described in detail in the Appendix.

5.1   File utility.dat contains monthly records of utility costs.

    a)   Generate descriptive statistics for telephone costs and fuel costs using the date variable to identify the extremes in the data set. Use LABEL and TITLE statements. Describe the shape of each distribution. Would you use percentiles or the sample mean and standard deviation to describe the shape of each distribution? Which dates have extreme values? Are they outliers?

    b)   Do part (a) for electricity costs and total costs.

5.2   Repeat Problem 5.1(b) for each year in the data set for the variable total costs.

5.3   File well#1.dat contains nitrate, zinc, and TDS (total dissolved solids) readings for a well being monitored for contamination.

    a)   Generate descriptive statistics for these variables. Use LABEL and TITLE statements. On what dates do outliers appear?

    b)   Change mg/l to g/l for TDS, and generate descriptive statistics. Does the shape of the stem-and-leaf plot change? How do the sample mean and sample standard deviation change?

5.4   The file china#1.dat contains import and export information for China. Generate descriptive statistics for the amount of imports, exports, and trade balance (exports – imports) using year as the ID variable. Use TITLE statements. Describe these distributions, including any information about outliers.

YOU SHOULD NOW KNOW

how to use PROC UNIVARIATE to generate descriptive statistics and plots

how to read the output and summarize the information SAS provides

how to make your SAS output easier to read using LABEL and TITLE statements

---

PROC Chart
MODULES NEEDED   1, 2, 5

Although the stem-and-leaf plot or histogram produced by PROC UNIVARIATE is helpful for looking at the shape of a distribution, you may want to have a larger or more sophisticated plot to look at. PROC CHART provides several options for looking at data.

PROC CHART

PROC CHART can create many different types of charts. VBAR creates a vertical bar chart. HBAR creates a horizontal bar chart.

PROC CHART will automatically choose the intervals and thus the number of bars in the bar charts unless you specify how many bars you want or what the midpoints of the intervals should be.

Bars can be "shaded" by denoting a SUBGROUP variable. Each value of the SUBGROUP variable creates a different "shading" by the use of a different character. The bars then reflect how much of the area is represented by the different values of the SUBGROUP variable.

PROC CHART General Form

```
proc chart data=data set;
   by variables;
   vbar variables/ <options>;
   hbar variables / <options>;
```

VBAR and HBAR Option	Description
`levels = # midpoints`	Specifies the number of bars on the chart. If you don't specify, SAS chooses the number of levels and the midpoints. Do not use with `midpoints`.
`midpoints = list`	When the variable is continuous, you can specify values for the midpoints. Do not use with `levels`. For example, these statements are valid: `midpoints = 10 20 30` `midpoints = 10 to 100 by 10`
`subgroup = variable`	The bar can be divided into parts representing the values of the specified variable. The first character of variable is used.
`type = freq` `type = pct`	The chart contains frequencies (the default) or percents.

EXAMPLE 6.1

In Example 5.1, descriptive statistics were generated for the variable patient improvement in the skin study. What does a histogram of the data look like? In this example, each bar of the histogram will show how much of the data is from the control group (1) and the drug group (2). The output is in Figure 6.1.

SAS Program

```
filename datain 'sclero.dat';

data one;
   infile datain;
   input @1 clinic 2. @4 id 2. @8 drug 1. @11 thick1 2.
      @15 thick2 2. @19 mobilty1 3. @24 mobilty2 3.
      @29 assess1 1. @32 assess21. ;
   improve = assess1 - assess2;
   label id      = 'Patient ID Number'
         drug    = 'Drug (1) or Placebo (2)'
         thick1  = 'Skin Thickening at 1st Visit'
         thick2  = 'Skin Thickening at 2nd Visit'
         mobilty1= 'Skin Mobility at 1st Visit'
         mobilty2= 'Skin Mobility at 2nd Visit'
         assess1 = 'Patient Assessment at 1st Visit'
         assess2 = 'Patient Assessment at 2nd Visit'
         improve = 'Improvement in Assessment Score'
         ;
   run;

   proc chart;
```

```
   vbar improve / subgroup = drug;
 title1 'Histogram of Patient Improvement in Assessment Score';
 run;
```

New Statements

```
proc chart;
vbar improve/ subgroup = drug;
```
Since neither LEVELS nor MIDPOINTS was used, the value under each bar is the midpoint of the interval that SAS chose. Each bar will reflect how many observations were due to the people receiving the treatment (DRUG=1) and how many were due to people receiving the placebo (DRUG=2).

Discussion of Output

Look at Figure 6.1 and notice that the bars are made up of 1s and 2s. The group receiving the drug is represented by 1s, and the control group is represented by 2s. Look at the third bar. The midpoint of the interval is −0.75. Eleven observations make up this bar, three for the treatment group and eight for the control group.

Figure 6.1   SAS Output for Example 6.1

```
     Histogram of Patient Improvement in Assessment Score    1
                             12:32 Monday, May 17, 1993


                    FREQUENCY OF IMPROVE

FREQUENCY

      |                        22222
      |                        22222
  35 +                         22222
      |                        22222
      |                        22222
      |                        22222
      |                        22222
  30 +                         22222
      |                        22222
      |                        22222
      |                        22222
      |                        22222
  25 +                         22222
      |                        22222
      |                        22222
      |                        22222
      |                        22222
  20 +                         22222
      |                        22222
      |                        22222
      |                        11111
      |                        11111
  15 +                         11111
      |                        11111
      |                        11111
      |                        11111
      |              22222      11111                22222
  10 +              22222      11111                22222
      |              22222      11111      22222      22222
      |              22222      11111      22222      22222
      |              22222      11111      22222      22222
      |              22222      11111      22222      22222
   5 +      22222    22222      11111      22222      11111
      |      22222    22222      11111      22222      11111
      |      22222    11111      11111      22222      11111
      |      11111    11111      11111      11111      11111
      | 11111 11111    11111      11111      11111      11111 11111
      ---------------------------------------------------------------
        -3.75   -2.25   -0.75    0.75     2.25    3.75    5.25

                    Improvement in Assessment Score


              SYMBOL DRUG      SYMBOL DRUG

                  1   1           2   2
```

PROBLEMS

The files for these problems are described in detail in the Appendix.

6.1   File utility.dat contains monthly records of utility costs. Generate histograms for telephone costs, fuel costs, electricity costs, and total costs. Use LABEL and TITLE statements. Describe the shape of these distributions.

6.2   Use the file utility.dat to see if the histograms differ by year for the years 1990–1992 for the variable

    a)   fuel costs. Use 25, 50, 75, 100, 125, and 150 for the midpoints.
    b)   total costs. Use 100, 140, 180, 220, 260, and 300 for the midpoints.
    c)   phone costs. Use 40, 60, 80, 100, 120, and 140 for the midpoints.

6.3   Generate histograms for nitrate, zinc, and TDS (total dissolved solids) from well#1.dat, and comment on the shape of the distribution. Use LABEL and TITLE statements.

6.4   The file china#1.dat contains import and export information for China.

    a)   Generate histograms for imports, exports, and trade balance (exports – imports), and comment on the shape of the distribution. Use LABEL and TITLE statements.
    b)   Generate a histogram for exports that has five levels or bars.
    c)   Generate a histogram for imports that has six levels or bars.

YOU SHOULD NOW KNOW

how to use PROC CHART to generate histograms

Descriptive Statistics II
MODULES NEEDED 1, 2, 5

Other procedures besides PROC UNIVARIATE generate descriptive statistics. PROC MEANS outputs the basic descriptive statistics in a more concise way than UNIVARIATE. When data are in categories, PROC FREQ can be used to generate a table containing counts of the variable(s) of interest, the overall percent, the row percent, and the column percent.

In this module, you will learn about PROC MEANS and PROC FREQ.

PROC MEANS

PROC MEANS is a good procedure to use when you are interested in the basic descriptive statistics. By default, it generates the sample size, the mean, the standard deviation, and the minimum and maximum values. It does not have the capability to generate the plots, percentiles, and largest and smallest values that PROC UNIVARIATE produces.

PROC MEANS General Form

```
proc means data=data set <options>;
    by variables;
    var variables;
```

MEANS Option	Description
n min max mean std	These are the default values if none are specified. They are, respectively, the sample size, the minimum value, the maximum value, the mean, and the standard deviation.
nmiss range sum var stderr t prt	These can also be specified and refer to, respectively, the number of missing observations, the range, the sum, the variance, the standard error of the mean, a t-statistic for testing whether the mean is significantly different from 0, and the p-value for that test.

PROC FREQ

PROC FREQ generates tables for data that are in categories. For one variable, a one-way table summarizes all the values of the variable, how many variables have each value, and the percent and cumulative percent for each value.

For two variables, a two-way table contains cell frequencies, cell percent of total, cell percent of row total, and cell percent of column total.

PROC FREQ uses a TABLES statement instead of a VAR statement when requesting tables. It is possible to request more than one table in a single TABLES statement.

PROC FREQ General Form

```
proc freq data=data set;
   by variables;
   tables var1*var2/<options>;
   weight variable;
```

TABLES Option	Description
chisq	Computes the chi-square statistic for testing for independence or homogeneity in two-way tables.
expected	Computes the expected counts for two-way tables.
nocol	Omits the column percents from the table.
nocum	Omits the cumulative frequencies for one-way tables.
nofreq	Omits the cell frequencies from the table.
nopercent	Omits any percents from the table.
norow	Omits the row percents from the table.

For a description of the WEIGHT statement, see Module 19.

EXAMPLE 7.1

Ms. Cohen wants summary statistics for her college statistics class.

- She wants to compute a final course grade based on weighting the total quiz grade 10%, the two midterm exams 20% each, the lab grade 10%, and the final exam 40%. There were a possible 50 points total for the quizzes, 100 for each midterm and the lab, and 200 for the final exam. (Because students scored lower than expected, she wants to adjust the scores up by 10%.) Then, she wants to change the percents to letter grades.
- She would also like to see how many students got As, how many got Bs, and so on.
- She is curious to see if there is a difference in grade distribution depending on the student's year in school.
- And she would like to check to see if people who did well on the quizzes (scored above 90%) did well on the midterm exams (scored above 80%) and whether this was different for women and men.

She will use PROC MEANS to compute summary statistics for the numerical grades and PROC FREQ to generate the table information. Figure 7.1 shows output from PROC MEANS, and Figures 7.2 and 7.3 show output from PROC FREQ.

SAS Program

```
filename indata 'grades.dat';

data one;
   infile indata;
   input id $ gender $ year quiz exam1 exam2 lab finexam;

   quiz = quiz*2;
   finexam = finexam/2 + 10;
   course = .1*quiz + .2*(exam1+exam2) + .1*lab + .4*finexam;
   course = course/100;

   if course >=.90 then coursltr = 'A';
   else if .80 <=course < .90 then coursltr = 'B';
   else if .70 <=course < .80 then coursltr = 'C';
   else if .60 <=course < .70 then coursltr = 'D';
   else if        course < .60 then coursltr = 'F';

   if quiz > 90 and (exam1 > 80 or exam2 > 80)
       then outcome = 'expected   ';
       else outcome = 'unexpected';

run;

proc means;
   var course quiz exam1 exam2 lab finexam;
title 'Summary Statistics from Statistics Class';
run;

proc freq;
   tables coursltr coursltr*year gender*year outcome*gender;
title 'Grades from Statistics Class';
run;
```

New Statements

```
proc means;
var course quiz exam1 exam2 lab finexam;
```
     Since no options were specified, PROC MEANS will generate the default statistics—number of observations, mean, standard deviation, minimum, and maximum.

```
proc freq;
tables coursltr coursltr*year gender*year outcome*gender;
```
Four tables are requested: a one-way table for the variable COURSLTR and two-way tables for COURSLTR by YEAR, GENDER by YEAR, and OUTCOME by GENDER.

Discussion of Output

In Figure 7.1, the average course grade was a B (mean=.84), students on average did better on the second midterm exam (exam1 mean=80.53 and exam2 mean=90.57), and the range of final exam grades was 47 to 105.5 after adjusting the scores.

In Figure 7.2, the course grade distribution was 15 As, 21 Bs, 6 Cs, 5 Ds, and 1 F. The percentage of students receiving a C or better was 85.7%.

Figure 7.3 shows the output for one of the tables, requested in the PROC FREQ TABLES statement—OUTCOME by GENDER. Eight is the number of observations where OUTCOME='expected' and GENDER='f'; 16.33 is the percent of total (8/49); 47.06 is the row percent (8/17); and 47.06 is the column percent (8/17). You can also see there are 17 observations where OUTCOME='expected' for a row percent of 34.69 (17/49). For the variable GENDER, 32 observations have a value of 'm' or 65.3% (32/49).

Figure 7.1   PROC MEANS Output

```
                  Summary Statistics from Statistics Class          1
                       14:55 Monday, May 17, 1993

 N Obs  Variable    N       Minimum        Maximum          Mean
 ------------------------------------------------------------------
    49  COURSE     49     0.5590000      1.0040000      0.8404286
        QUIZ       49    40.0000000    100.0000000     85.6734694
        EXAM1      49    33.0000000    100.0000000     80.5306122
        EXAM2      49    51.0000000    100.0000000     90.5714286
        LAB        49    63.0000000    100.0000000     93.5306122
        FINEXAM    49    47.0000000    105.5000000     79.7551020
 ------------------------------------------------------------------


        N Obs  Variable      Variance        Std Dev
        --------------------------------------------------
           49  COURSE        0.0111601      0.1056415
               QUIZ        172.2244898     13.1234329
               EXAM1       309.2542517     17.5856263
               EXAM2       103.7083333     10.1837289
               LAB          46.4209184      6.8132898
               FINEXAM     192.3450255     13.8688509
        --------------------------------------------------
```

Figure 7.2   PROC FREQ Output for COURSLTR

```
                Grades from Statistics Class                    2
                           14:55 Monday, May 17, 1993

                                    Cumulative  Cumulative
      COURSLTR   Frequency   Percent  Frequency   Percent
      ---------------------------------------------------------
      A             15        30.6       15        30.6
      B             21        42.8       36        73.5
      C              6        12.3       42        85.7
      D              5        10.1       47        95.9
      F              2         4.1       49       100.0
```

Figure 7.3   PROC FREQ Output for OUTCOME by GENDER

```
        Grades from Statistics Class                    3
              14:55, Monday, May 17, 1993

                  TABLE OF OUTCOME BY GENDER

        OUTCOME      GENDER

        Frequency |
        Percent   |
        Row Pct   |
        Col Pct   |f        |m        | Total
        ----------+---------+---------+
        expected  |     8 |      9 |    17
                  | 16.33 | 18.37 | 34.69
                  | 47.06 | 52.94 |
                  | 47.06 | 28.13 |
        ----------+---------+---------+
        unexpected|     9 |     23 |    32
                  | 18.37 | 46.94 | 65.31
                  | 28.13 | 71.88 |
                  | 52.94 | 71.88 |
        ----------+---------+---------+
        Total          17       32       49
                     34.69    65.31   100.00
```

PROBLEMS

The files for these problems are described in detail in the Appendix.

7.1   Use the file utility.dat and produce summary statistics for all numeric variables using PROC MEANS, including one for total costs.

7.2   Do Problem 7.1 for each year.

7.3   Use PROC MEANS to produce summary statistics for total dissolved solids (TDS) in file well#15.dat.

7.4   Generate summary statistics for zinc concentrations in file well#8.dat.

7.5   Use the file well#1.dat, which contains data on nitrate, zinc, and TDS concentrations.

a)   Create a new variable to reflect that the well is contaminated if nitrate > 0.12, zinc > 0.02, or TDS > 516. If none of these conditions hold, the well is not contaminated. Find out how many readings indicate the well is contaminated by using PROC FREQ.

b)   Create a two-way table for contamination by year. Do you notice anything unusual?

7.6   The file survresp.dat contains response rate information from mail surveys using different monetary incentives. Create a new variable that indicates a successful outcome if the response rate from the control group is lower than the response rate from the treatment group. Use a LABEL statement. Generate a one-way table for outcome and a two-way table for outcome and the different incentives. What do you conclude about the use of incentives in mail surveys?

7.7   Use the scleroderma study data in sclero.dat. Create a new variable that indicates a good result if skin thickening decreased, another one to indicate a good result if skin mobility increased, and a third one to indicate a good result if patient assessment decreased. Assign a LABEL to all the improvement variables.

a)   Which clinic(s) had the largest number of patients in the study?

b)   What percent of subjects from each clinic were in the control group (took the placebo)?

c)   For clinics 46, 48, and 49, generate two-way tables for drug/placebo and the percent whose skin thickening improved, the percent whose skin mobility improved, and the percent whose patient assessment improved.

d)   Generate a two-way table that looks at skin thickening improvement and patient assessment improvement for those who took the drug (not the placebo). What percent had bad outcomes in both measures? What percent had good outcomes in both measures?

YOU SHOULD NOW KNOW

how to generate descriptive statistics using PROC MEANS

how to generate one-way and two-way tables using PROC FREQ

Generating Random Observations

MODULES NEEDED 1, 2, 5, 6

What do the distributions of random variables look like? There are functions in SAS that can be used to generate observations from specific distributions. Generating random observations in SAS involves DO loops. This is a way to repeat a command or statement over and over again. One can then use PROC CHART to look at the shape of the distribution.

In this module, you will learn how to generate observations from certain distributions using SAS distributional functions and DO loops.

SAS DISTRIBUTIONAL FUNCTIONS

With SAS you can generate random observations from discrete and continuous distributions such as binomial, Poisson, geometric, hypergeometric, normal, exponential, and uniform. This is done with the use of a function and a seed. The function indicates which distribution you are interested in, and the seed is a number used by the random number generating algorithm. The seed is used to start the algorithm that generates random numbers. The first seed SAS encounters is the one that begins the algorithm: Subsequent seeds in other statements are ignored. The seed can be any positive number.

Here is a list of some functions and a description of the random observation that is generated.

SAS Function	Description: A Random Observation from a(n)
ranbin(seed,n,p)	binomial distribution with n trials and probability p of success
ranexp(seed)	exponential distribution with $\lambda = 1$
rannor(seed)	standard normal distribution ($\mu = 0$ and $\sigma = 1$)
ranpoi(seed,m)	Poisson distribution with mean $m \geq 0$
ranuni(seed)	uniform distribution on the interval (0,1)

Although the functions generate random numbers from a "standard" distribution, it is possible to make a change of variables to generate numbers from other distributions in that family. If X is a normal random variable with mean μ and standard deviation σ and Z is a standard normal random variable, then $X = Z*\sigma + \mu$. If X is an exponential random variable with $\lambda = 7$ and Y is an exponential random variable with $\lambda = 1$, then $X = Y \div 7$. If X is a uniform distribution on the interval (a,b) and Y is a uniform distribution on (0,1), then X = (b–a)*Y + a.

DO LOOPS

DO loops are used when a statement or set of statements needs to be repeated many times. DO loops begin with a DO statement and end with an END statement. The statements in between are repeated the number of times indicated by the index variable in the DO statement.

General Form of a DO Loop

```
do index variable = beginning number to ending number;
  statements
end;
```

EXAMPLE 8.1

Professor Mason wants to generate 1000 observations from a standard normal distribution and then generate a histogram of the data. See Figure 8.1 for the log file and Figure 8.2 for the output file.

SAS Program

```
data one;
   do i = 1 to 1000;
      x = rannor(46327);
      output;
   end;
run;

proc chart;
   vbar x/ midpoints = -3 to 3 by .5;
title 'Random Observations from Standard Normal Distribution';
run;
```

New Statements

```
do i = 1 to 1000;
x = rannor(46327);
output;
```

`end;`   This is a DO loop. This DO loop builds a data set that has two variables, I and X, and it will have 1000 observations because the second and third statements are repeated 1000 times.

`x = rannor(46327);`

RANNOR generates a standard normal observation using 46327 as the seed to start the algorithm.

`output;`   This statement says OUTPUT the observation to DATA ONE. It is an implied statement at the end of every DATA step. Here, it is necessary to state it explicitly. Without the OUTPUT statement, the DO loop would just evaluate X 1000 times and stop. There would be no observations in the data set.

Discussion of Output

In Figure 8.1, the first NOTE in the SAS log says that the data set has 1000 observations and 2 variables. The two variables are X and I.

In Figure 8.2, you can see a histogram of the 1000 randomly generated observations. Notice the bell curve shape.

Figure 8.1   SAS Log for Example 8.1

```
  1      data one;
  2        do i = 1 to 1000;
  3          x = rannor(46327);
  4          output;
  5        end;
  6      run;
NOTE: The data set WORK.ONE has 1000 observations and
      2 variables.
NOTE: The DATA statement used 4.00 seconds.
  7
  8      proc chart;
  9        vbar x / midpoints = -3 to 3 by .5;
 10      title 'Random Observations from Standard Normal Distribution';
 11      run;
NOTE: The PROCEDURE CHART used 4.00 seconds.
```

Figure 8.2   SAS Output for Example 8.1

```
      Random Observations from Standard Normal Distribution
                                    13:02 Wednesday April 28, 1993

Frequency

        |                          ***
        |                  ***     ***
  180 + |                  ***     ***
        |                  ***     ***     ***
        |                  ***     ***     ***
        |                  ***     ***     ***
  160 + |                  ***     ***     ***
        |                  ***     ***     ***
        |                  ***     ***     ***
        |                  ***     ***     ***
  140 + |                  ***     ***     ***
        |                  ***     ***     ***
        |                  ***     ***     ***
        |                  ***     ***     ***
  120 + |          ***     ***     ***     ***
        |          ***     ***     ***     ***     ***
        |          ***     ***     ***     ***     ***
        |          ***     ***     ***     ***     ***
  100 + |          ***     ***     ***     ***     ***
        |          ***     ***     ***     ***     ***
        |          ***     ***     ***     ***     ***
        |          ***     ***     ***     ***     ***
   80 + |          ***     ***     ***     ***     ***
        |      ***     ***     ***     ***     ***     ***
        |      ***     ***     ***     ***     ***     ***     ***
        |      ***     ***     ***     ***     ***     ***     ***
   60 + |      ***     ***     ***     ***     ***     ***     ***
        |      ***     ***     ***     ***     ***     ***     ***
        |      ***     ***     ***     ***     ***     ***     ***
        |      ***     ***     ***     ***     ***     ***     ***
   40 + |      ***     ***     ***     ***     ***     ***     ***
        |      ***     ***     ***     ***     ***     ***     ***
        |  ***     ***     ***     ***     ***     ***     ***     ***
        |  ***     ***     ***     ***     ***     ***     ***     ***     ***
   20 + |  ***     ***     ***     ***     ***     ***     ***     ***     ***
        |  ***     ***     ***     ***     ***     ***     ***     ***     ***
        | ***  ***     ***     ***     ***     ***     ***     ***     ***     ***
        | ***  ***  ***     ***     ***     ***     ***     ***     ***     ***   ***   ***
        -------------------------------------------------------------------------
           -3.0 -2.5 -2.0 -1.5 -1.0 -0.5  0.0  0.5  1.0  1.5  2.0  2.5  3.0

                                   X Midpoint
```

EXAMPLE 8.2

Professor Mason now wants to generate 1000 observations from a uniform distribution on the interval (10, 60) and create a histogram. Refer to Figure 8.3 for the output.

SAS Program

```
data one;
   do i = 1 to 1000;
     uni_0_1 = ranuni(55223);
     uni_1060 = uni_0_1*50 + 10;
     output;
   end;
run;

proc chart data=one;
   vbar uni_1060 / levels=10;
title 'Random Observations from Uniform (10,60) Distribution';
run;
```

New Statements

```
uni_0_1 = ranuni(55223);
uni_1060 = uni_0_1*50 + 10;
```

            The new variable UNI_0_1 has a value from a uniform distribution on (0,1). To change to a uniform random variable on (10, 60), we need to change the scale. That is done by multiplying UNI_0_1 by 50 and adding 10.

Discussion of Output

Figure 8.3 shows the distribution of 1000 observations from a uniform distribution on the interval (10, 60). SAS chose the midpoints of the bars. The first bar has midpoint 12.5; it includes observations on the interval (10, 15). Notice that the height of the bars does not vary greatly from bar to bar. The heights are not uniform, however, because the observations reflect a random sample of values.

Figure 8.3   SAS Output from Example 8.2

```
               Random Observations from Uniform (10,60) Distribution
                                         13:02 Wednesday, April 28, 1993

Frequency

       |                                          *****
  110 +|                                          *****
       |                         *****            *****
       |                         *****            *****
       |                         *****            *****                      *****
  100 +|           *****         *****            *****            *****     *****
       |   *****   *****         *****            *****   *****    *****     *****
       |   *****   *****         *****   *****    *****   *****    *****   *****   *****
       |   *****   *****         *****   *****    *****   *****    *****   *****   *****
   90 +|   *****   *****  *****  *****   *****    *****   *****    *****   *****   *****
       |   *****   *****  *****  *****   *****    *****   *****    *****   *****   *****
       |   *****   *****  *****  *****   *****    *****   *****    *****   *****   *****
       |   *****   *****  *****  *****   *****    *****   *****    *****   *****   *****
   80 +|   *****   *****  *****  *****   *****    *****   *****    *****   *****   *****
       |   *****   *****  *****  *****   *****    *****   *****    *****   *****   *****
       |   *****   *****  *****  *****   *****    *****   *****    *****   *****   *****
       |   *****   *****  *****  *****   *****    *****   *****    *****   *****   *****
   70 +|   *****   *****  *****  *****   *****    *****   *****    *****   *****   *****
       |   *****   *****  *****  *****   *****    *****   *****    *****   *****   *****
       |   *****   *****  *****  *****   *****    *****   *****    *****   *****   *****
       |   *****   *****  *****  *****   *****    *****   *****    *****   *****   *****
   60 +|   *****   *****  *****  *****   *****    *****   *****    *****   *****   *****
       |   *****   *****  *****  *****   *****    *****   *****    *****   *****   *****
       |   *****   *****  *****  *****   *****    *****   *****    *****   *****   *****
       |   *****   *****  *****  *****   *****    *****   *****    *****   *****   *****
   50 +|   *****   *****  *****  *****   *****    *****   *****    *****   *****   *****
       |   *****   *****  *****  *****   *****    *****   *****    *****   *****   *****
       |   *****   *****  *****  *****   *****    *****   *****    *****   *****   *****
       |   *****   *****  *****  *****   *****    *****   *****    *****   *****   *****
   40 +|   *****   *****  *****  *****   *****    *****   *****    *****   *****   *****
       |   *****   *****  *****  *****   *****    *****   *****    *****   *****   *****
       |   *****   *****  *****  *****   *****    *****   *****    *****   *****   *****
       |   *****   *****  *****  *****   *****    *****   *****    *****   *****   *****
   30 +|   *****   *****  *****  *****   *****    *****   *****    *****   *****   *****
       |   *****   *****  *****  *****   *****    *****   *****    *****   *****   *****
       |   *****   *****  *****  *****   *****    *****   *****    *****   *****   *****
       |   *****   *****  *****  *****   *****    *****   *****    *****   *****   *****
   20 +|   *****   *****  *****  *****   *****    *****   *****    *****   *****   *****
       |   *****   *****  *****  *****   *****    *****   *****    *****   *****   *****
       |   *****   *****  *****  *****   *****    *****   *****    *****   *****   *****
       |   *****   *****  *****  *****   *****    *****   *****    *****   *****   *****
   10 +|   *****   *****  *****  *****   *****    *****   *****    *****   *****   *****
       |   *****   *****  *****  *****   *****    *****   *****    *****   *****   *****
       |   *****   *****  *****  *****   *****    *****   *****    *****   *****   *****
       |   *****   *****  *****  *****   *****    *****   *****    *****   *****   *****
       ----------------------------------------------------------------------------------
            12.5    17.5    22.5    27.5    32.5    37.5    42.5    47.5    52.5    57.5
                                    UNI_1060 Midpoint
```

EXAMPLE 8.3

Professor Maestas wants to look at the shape of the distribution of a sample mean. She knows that if the sample size is "large," the Central Limit Theorem states that the sample mean is approximately normally distributed. She decides to generate 500 samples each having 15 observations from a binomial distribution with $p=0.3$ and $n=50$. For each of the 500 samples, she will compute the mean of those 15 observations. The mean represents the average number of successes from 50 trials with the probability of a success being 0.3. Observations from such a Binomial distribution will be approximately normal with a mean of 15 and a variance of 10.5. The average of 15 of these observations will also be approximately normal with a mean of 15 and a variance of 10.5/15. See Figure 8.4 for the output.

SAS Program

```
data one;
   do i = 1 to 500;
      x1 = ranbin(4491,50,.3);
      x2 = ranbin(4491,50,.3);
      x3 = ranbin(4491,50,.3);
      x4 = ranbin(4491,50,.3);
      x5 = ranbin(4491,50,.3);
      x6 = ranbin(4491,50,.3);
      x7 = ranbin(4491,50,.3);
      x8 = ranbin(4491,50,.3);
      x9 = ranbin(4491,50,.3);
      x10 = ranbin(4491,50,.3);
      x11 = ranbin(4491,50,.3);
      x12 = ranbin(4491,50,.3);
      x13 = ranbin(4491,50,.3);
      x14 = ranbin(4491,50,.3);
      x15 = ranbin(4491,50,.3);
      ave_x = (x1+x2+x3+x4+x5+x6+x7+x8+x9+x10+x11+x12+x13+
              x14+x15)/15;
      output;
   end;
run;

proc chart;
   vbar ave_x;
title1 'Distribution of the Mean of 500 Samples';
title2 'from a Binomial (50, 0.3) Distribution';
run;
```

New Statements

```
do i = 1 to 500;
   x1 = ranbin(4491,50,.3);
```

```
x2 = ranbin(4491,50,.3);
x3 = ranbin(4491,50,.3);
⋮
x14 = ranbin(4491,50,.3);
x15 = ranbin(4491,50,.3);
ave_x = (x1+x2+x3+x4+x5+x6+x7+x8+x9+x10+x11+x12+x13+
          x14+x15)/15;
output;
end;
```

When I=1, 15 variables named X1 – X15 are generated using the function RANBIN. The seed is the same because SAS uses the seed only the first time it encounters a random number generator. AVE_X is the average of the 15 observations. This constitutes one of the 500 simulated samples. When I=2, another 15 observations are generated and their average computed. This continues 500 times. The 500 values of AVE_X will be used to construct a histogram with PROC CHART.

Discussion of Output

In Figure 8.4, the distribution of the mean values calculated from the 500 samples with n=15 looks somewhat bell-shaped. Probability theory says it should be centered on 15 with a spread of ± 2.5 (3 times the standard deviation 0.8367). Notice that this is the case.

PROBLEMS

8.1   Generate 1000 observations from a normal distribution with a mean of 50 and a standard deviation of 20. Use PROC CHART to look at this distribution. Describe the shape of the distribution.

8.2   Generate the following number of observations from a normal distribution with a mean of 10 and a standard deviation of 10. Use PROC CHART to look at this distribution. Describe the shape of the distribution.

    a)   50 observations
    b)   500 observations
    c)   5000 observations
    d)   What do you notice about the shapes of the distributions in (a), (b), and (c)?

8.3   Generate 1000 observations from an exponential distribution with $\lambda=7$. Create a histogram of this distribution. Describe the shape of the distribution.

Figure 8.4   SAS Output for Example 8.3

```
                    Distribution of the Mean of 500 Samples
                      from a Binomial (50, 0.3) Distribution
                                      13:02 Wednesday, April 28, 1993

   Frequency

         |                                   ****
         |                                   ****
    120 +                            ****    ****
         |                            ****    ****
         |                            ****    ****
         |                            ****    ****
    100 +                            ****    ****
         |                            ****    ****
         |                    ****    ****    ****
         |                    ****    ****    ****
     80 +                    ****    ****    ****
         |                    ****    ****    ****
         |                    ****    ****    ****
         |                    ****    ****    ****
     60 +                    ****    ****    ****    ****
         |                    ****    ****    ****    ****
         |                    ****    ****    ****    ****
         |                    ****    ****    ****    ****    ****
     40 +                    ****    ****    ****    ****    ****
         |                    ****    ****    ****    ****    ****
         |            ****    ****    ****    ****    ****    ****
         |            ****    ****    ****    ****    ****    ****
     20 +            ****    ****    ****    ****    ****    ****
         |            ****    ****    ****    ****    ****    ****
         |    ****    ****    ****    ****    ****    ****    ****    ****
         |    ****    ****    ****    ****    ****    ****    ****    ****    ****
         ---------------------------------------------------------------------------
           12.75 13.25 13.75 14.25 14.75 15.25 15.75 16.25 16.75 17.25 17.75

                              sample average - n=15
```

8.4   Generate 700 observations from a Poisson distribution with a mean of 5, and create a histogram of these values. Describe the shape of the distribution.

8.5   Generate 500 observations from a binomial distribution with n=40 and p=0.2. Create a histogram. Describe the shape of the distribution.

8.6   Generate 1000 random samples of size 10 from an exponential distribution with $\lambda=7$. Create a new variable to find the average of these ten values. Use PROC CHART to look at the distribution of the mean. Describe the shape of the distribution.

8.7   Generate 1000 random samples of size 10 from a uniform distribution on (10, 20). Create a new variable to find the average of these ten values. Use PROC CHART to look at the distribution of the mean. Describe the shape of the distribution.

YOU SHOULD NOW KNOW

how to use a DO loop

how to generate random observations

X–Y Plots

MODULES NEEDED 1, 2, 5

When data are collected in pairs, the two variables may be related to each other in some way. One often wants to plot the observations on a graph in order to see if a relationship exists. PROC PLOT is used to generate X-Y plots.

In this module, you will use PROC PLOT to plot data.

PROC PLOT

PROC PLOT generates X-Y plots using a PLOT statement to specify which variable is to be on the X-axis and which is to be on the Y-axis. Multiple plots can be requested with one PLOT statement. Plots can appear on separate pages or overlayed on the same page. Different characters can be used for the plotting symbol.

When two points are very close together, SAS may not be able to show both of them on the plot. When that happens, a note appears on the plot that some observations are hidden.

PROC PLOT General Form

```
proc plot data=data set;
  by variables;
  plot yvar * xvar;
```

PLOT Statement Variation	Description
yvar * xvar = 'char'	Observations are plotted using the character specified, such as '+' or 'q' or '3'.
yvar * xvar = variable	The first character of the value of variable is plotted.
yvar * (xvar1 xvar2)	Two plots yvar*xvar1 and yvar*xvar2 appear on separate pages.
yvar1 * xvar1 = 'char1' yvar2 * xvar2 = 'char2' / overlay	Two plots yvar1*xvar1 and yvar2*xvar2 appear on the same plot. They are overlayed. Plotting characters distinguish between the two plots.

EXAMPLE 9.1

Mr. Chang is interested in looking at China's imports over time. He wants to create a plot with imports on the Y-axis and year on the X-axis. (A complete description of this data is in the Appendix.) The output is shown in Figure 9.1.

SAS Program

```
filename in 'china#1.dat';

data trade;
   infile in;
   input year 1-4 total 6-10 export 12-16 import 18-22;
run;

proc plot;
   plot import*year = '+';
title 'Plot of China''s Imports for Years 1955-89';
run;
```

New Statements

```
proc plot;
plot import*year='+';
```
> The variable IMPORT will be on the Y-axis, and the variable YEAR will be on the X-axis. The plotting symbol is a plus sign (+).

Discussion of Output

China's imports did not change much from 1955 to 1970. After that time, the imports increased dramatically.

EXAMPLE 9.2

Mr. Chang is interested in looking at China's imports and exports during the 1980s. He wants to create a plot using different symbols for imports and exports on the Y-axis and year on the X-axis. The output is shown in Figure 9.2.

SAS Program

```
filename in 'china#1.dat';

data trade;
   infile in;
```

Figure 9.1   PROC PLOT Output from Example 9.1

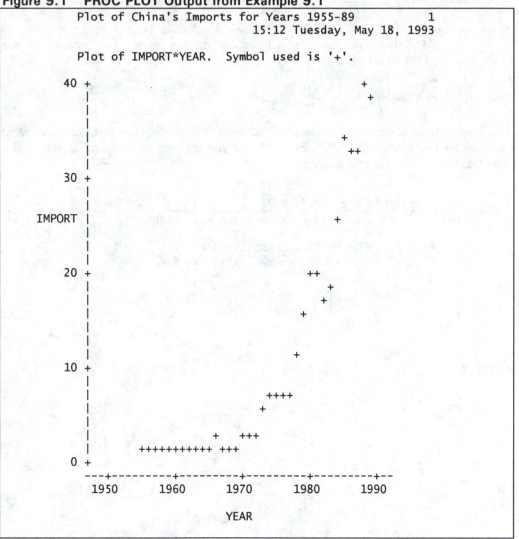

```
              Plot of China's Imports for Years 1955-89           1
                            15:12 Tuesday, May 18, 1993


        Plot of IMPORT*YEAR.  Symbol used is '+'.


     40 +                           .                    +
        |                                                +
        |
        |
        |                                           +
        |                                          ++
        |
     30 +
        |
 IMPORT |                                      +
        |
        |
     20 +                              ++
        |                               +
        |                              +
        |                            +
        |
        |
     10 +                         +
        |
        |                      ++++
        |                     +
        |
        |              +    +++
        |        ++++++++++++ +++
      0 +
        ---+---------+---------+---------+---------+--
         1950      1960      1970      1980      1990

                          YEAR
```

```
    input @1 year 4. @6 total 5. @12 export 5. @18 import 5.;
    if 1980 <= year <= 1989;
run;

proc plot;
    plot import*year = '+' export*year = '#' / overlay;
title 'Plot of China''s Imports / Exports for Years 1980-89';
run;
```

New Statement

```
plot import*year = '+' export*year = '#' / overlay;
```
> The two plot requests will appear on the same plot because of the OVERLAY option.

Discussion of Output

China's imports and exports were close for the years 1980–1989. Note that the values were so close for the years 1981, 1984, and 1988 that you see only one plotting symbol and a note that three observations are hidden.

Figure 9.2   Output from Example 9.2

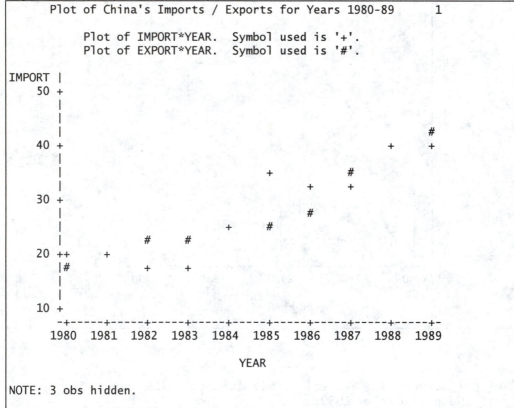

```
        Plot of China's Imports / Exports for Years 1980-89      1

          Plot of IMPORT*YEAR.  Symbol used is '+'.
          Plot of EXPORT*YEAR.  Symbol used is '#'.

IMPORT |
    50 +
       |
       |
       |                                                        #
    40 +                                              +         +
       |
       |                              +           #
       |                                  +       +
    30 +
       |                                      #
       |                          +       #
       |               #       #
    20 ++       +
       |#               +       +
       |
       |
    10 +
        -+-----+-----+-----+-----+-----+-----+-----+-----+-----+-
        1980  1981  1982  1983  1984  1985  1986  1987  1988  1989
                                  YEAR

NOTE: 3 obs hidden.
```

PROBLEMS

The files for these problems are described in detail in the Appendix.

9.1   Use utility.dat to create plots for the years 1989–1992 for the following variables over time. (HINT: You will need to change month to a numeric variable so that months will be in the correct order.) Describe the relationship between time and the other variables.

    a)   telephone costs
    b)   fuel costs
    c)   electricity costs
    d)   total costs

9.2   Redo Problem 9.1 overlaying phone, fuel, and electric costs for each year. Be sure to choose different plotting characters for each year. Do you notice any patterns?

9.3   Use well#8.dat to plot zinc concentrations over time. Describe the relationship between the two variables. (NOTE: In order to use the date value correctly, read in day and year as numeric variables, and change month to a numeric variable. Then use the following lines of code in the DATA step, which allows SAS to treat the date as a numeric, continuous variable: FORMAT DATE DATE7.; DATE = MDY(MO, DAY, YEAR);.)

9.4   Use handinj.dat to create an X-Y plot of days lost and cost. Choose something fun for the plotting character. Describe the relationship between the two variables.

9.5   Use survresp.dat to plot response rate improvement (= difference in rates / response rate for control group) as a function of incentive. Describe the relationship between the two variables.

YOU SHOULD NOW KNOW

how to generate X-Y plots

One-Sample Tests for μ, p

MODULES NEEDED 1, 2, 5, 7

You can use SAS procedures for hypothesis testing. PROC MEANS will do a one-sample t-test for the true mean of a distribution, and PROC FREQ will provide the information for you to do a test for the true proportion of a distribution.

In this module, you will learn how to do a one-sample t-test on a population mean using PROC MEANS and a test for the population proportion p using PROC FREQ.

ONE-SAMPLE T-TESTS

PROC MEANS will compute the t-statistic and p-value associated with $Ho:\mu=0$ vs. $Ha:\mu\neq0$. The option T requests the t-statistic, and the option PRT requests the p-value for the two-sided test. The general form of PROC MEANS is in Module 7.

If you are interested in a μ_0 value other than 0, you can recode the data by subtracting μ_0 from each observation. Testing that the true mean of the recoded data equals 0 is the same as testing whether the true mean of the original data equals μ_0.

If you want to do a one-sided test instead of a two-sided test, divide the p-value on the printout by two. The result is the p-value for a one-sided test.

EXAMPLE 10.1

Jamila Gandhi wants to test if the average TDS level in well #15 is greater than 975 at an α-level of 0.05. If it is, the well is contaminated, and she must begin corrective action to clean up the well. The output is shown in Figure 10.1.

SAS Program

```
filename inwell 'well#15.dat';

data one;
   infile inwell;
   input @1 date $char5. year tds;
   testtds = tds - 975;
run;

proc means n mean std t prt;
```

```
    var testtds;
title1 'Testing the Mean TDS Level for Well #15';
title2 'Is the True Mean > 975?';
run;
```

New Statements

```
testtds = tds - 975;
```
> The hypothesis is $H_0: \mu = 975$ vs. $H_a: \mu > 975$. Because PROC MEANS tests whether a mean *is equal to 0*, it is necessary to subtract 975 from each observation and run PROC MEANS for this new variable.

```
proc means n mean std t prt;
var testtds;
```
Options T and PRT generate the t-statistic and its p-value.

Discussion of Output

In Figure 10.1, the value of the t-statistic is 2.5501410, and the p-value for a two-sided test is 0.0176. Because the p-value is < 0.05, the null hypothesis would be rejected. The conclusion is that the true mean TDS level is greater than 975.

Figure 10.1 PROC MEANS Output for Example 10.1

```
              Testing the Mean TDS Level for Well #15           1
                     Is the True Mean > 975?
                               15:12 Tuesday, May 18, 1993

Analysis Variable : TESTTDS


N Obs    N         Mean        Std Dev          T  Prob>|T|
-----------------------------------------------------------------
  25    25    22.8800000    44.8602645    2.5501410    0.0176
-----------------------------------------------------------------
```

EXAMPLE 10.2

Jonathan Wright wants to know if the true proportion of people who show improvement as measured by the patient assessment score after taking the new drug tested in the skin study is greater than 0.50.

SAS Program

```
filename indata 'sclero.dat';

data arth;
```

```
   infile indata;
   input clinic id drug thick1 thick2 mobil1 mobil2
     assess1 assess2;
   if drug=1;
   if assess1 > assess2 then improve = 'yes';
   else if assess1 > assess2 then improve = 'no ';
run;

proc freq;
   tables improve;
title 'Proportion of Those Improving -- Assessment Score';
run;
```

Discussion of Output

From the output (see Figure 10.2), the sample proportion is 0.594. The test statistic is

$$z = \frac{.594 - .5}{\sqrt{\dfrac{.594 * (1 - .594)}{32}}} = 1.08$$

From a standard normal table, the p-value for this upper tailed test is 0.1401.   Since the p-value is large, the null hypothesis would not be rejected. The conclusion is that the proportion who show improvement is not greater than 0.50.

Figure 10.2 PROC FREQ Output for Example 10.2

```
        Proportion of Those Improving -- Assessment Score       1
                                 8:46 Thursday, May 20, 1993


                                    Cumulative  Cumulative
    IMPROVE  Frequency   Percent    Frequency    Percent
    -------------------------------------------------------
    no            13      40.6          13        40.6
    yes           19      59.4          32       100.0
```

PROBLEMS
The files for these problems are described in detail in the Appendix.

10.1   Use file well#1.dat to answer the following questions:

   a)   Is the true average nitrate concentration greater than 0.10?
   b)   Is the true average zinc concentration less than 0.01?
   c)   Is the true average TDS concentration different from 475?

10.2   Use file utility.dat to answer the following questions:

a)   Is the true average monthly phone cost higher than $50 per month?

b)   Is the true average monthly electric cost different from $30 per month?

10.3   Use file debate.dat to answer the following questions:

a)   Is the true proportion of students who believe that debate is more effective compared to other classes greater than 0.75?

b)   Redo (a) for Skyline High School.

c)   Is the true proportion of students who believe that the effectiveness of speech in teaching argumentation skills is very effective different from 0.80?

d)   Redo (c) for women.

e)   Is the true proportion of students who believe that the effectiveness of speech in teaching research skills is very effective less than 0.75?

f)   Is the true proportion of students who believe that the effectiveness of speech in teaching reasoning skills is very effective different from 0.95?

g)   Is the true proportion of male Taylorsville High School students who believe that the effectiveness of speech in teaching speaking skills is very effective less than 0.75?

10.4   Use file src.dat to answer the following questions:

a)   Is the true proportion of Utahns who identify strongly as environmentalists (respond with 8, 9, or 10) greater than 0.25?

b)   Is the true proportion of Utahns who strongly disagree or somewhat disagree that plants and animals exist primarily to be used by humans less than 0.45?

c)   Is the true proportion of Utahns who are working different from 0.50?

d)   Is the true proportion of Utahns who identify as conservative (answer 1 or 2) greater than 0.30?

YOU SHOULD NOW KNOW

how to use PROC MEANS to do a one-sample t-test

how to use PROC FREQ to generate the statistics to do a hypothesis test for a population proportion

Two-Sample t-Tests

MODULES NEEDED   1, 2, 5, 7

Whereas PROC MEANS is used to do a one-sample t-test, PROC TTEST is used to test whether two population means are equal or not. This module will discuss how to use PROC TTEST and PROC MEANS for t-tests—two-sample, pooled, and paired.

PROC TTEST

PROC TTEST tests whether two means are equal. It reports p-values for the case where the two variances are unequal, which is the two-sample test, and for the case where the two variances are equal, the pooled t-test. For the two-sample test, it computes the approximate degrees of freedom. PROC TTEST also reports the results of testing whether the two variances are equal for deciding which test—a pooled t-test or the two-sample t-test—is appropriate.

The CLASS statement identifies the variable that divides the data set into two groups. The CLASS variable must have only two values, and the values can be either numeric or character.

PROC TTEST General Form

```
proc ttest data=data set;
   by variables;
   class variable;
   var variables;
```

EXAMPLE 11.1

In a study to determine if center high-mounted stop lamps should be mandatory on trucks as they are now on cars, four truck types were used (pickup, cargo van, minivan, and straight truck). A researcher is interested in whether the control group (light was not on) and the experimental group differ regarding brake response time of the vehicle following the test truck. Specifically, the researcher wants the answer for two groups: cargo vans in 40 mph speed zone and straight trucks in a 50 mph speed zone. The output is in Figure 11.1.

SAS Program

```
filename inbrakes 'taillite.dat';

data one;
    infile inbrakes;
    input @1 id $2. @4 vehtype $1. @7 group $1. @10 positn $1.
        @13 speedzn 2. @17 resptime 3. @22 follotme 2.
        @28 folltmec $1.;

    label vehtype = 'Vehicle Type'
          group   = 'Group - Light On=1   Light Off=2'
          positn  = 'Light Position'
          speedzn = 'Speed Zone'
          resptime= 'Response Time'
          follotme= 'Following Time in Video Frames'
          folltmec= 'Following Time in Categories'
          ;
run;

data vans40;
    set one;
    if vehtype = 2 and speedzn = 40;
run;

data trucks50;
    set one;
    if vehtype = 4 and speedzn = 50;
run;

proc ttest data=vans40;
    class group;
    var resptime;
title1 'Comparing the Brake Response Time for';
title2 'Cargo Vans in 40 mph Speed Zones';
run;

proc ttest data=trucks50;
    class group;
    var resptime;
title2 'Straight Trucks in 50 mph Speed Zones';
run;
```

New Statements

```
proc ttest data=vans40;
class group;
var resptime;
```

The variable GROUP has two values 1 and 2 that divide the data set into two groups. The null hypothesis is that the mean response time for group 1 equals the mean response time for group 2.

Discussion of Output

For cargo vans in 40 mph speed zones, the F-statistic for testing whether the variances from the two groups are equal is 1.59 with a p-value of 0.1589. Because the p-value is large, one would conclude that the two variances are not unequal, continuing with a pooled t-test for the two means. The value of the pooled t is 0.1379, and the p-value is 0.8906. Clearly, the two means are not different.

For straight trucks in 50 mph speed zones, again a pooled t-test is appropriate (p-value for F is 0.1160). The pooled $t=-2.9371$, and the p-value$=0.0042$. One would reject the null hypothesis of equal means and conclude that the true average response times for the two groups are not equal.

T-TESTS: TWO-SAMPLE, POOLED, AND PAIRED

Use PROC TTEST for the two-sample t-test and the pooled t-test. The two-sample t-test assumes that $\sigma_1 \neq \sigma_2$. On the printout, the two-sample t is found on the line beginning with Unequal, referring to unequal variances. The pooled t-statistic is found on the line beginning with Equal.

For a paired t-test, use PROC MEANS. The paired t-test looks at the differences between two measures that are dependent or correlated and tests whether or not the mean difference equals zero. To use PROC MEANS for a paired t-test, create a new variable that is the difference between the two measures, and test whether the difference is equal to 0. See Module 7 for a description of PROC MEANS.

EXAMPLE 11.2

Ms. Cohen would like to know if the average score on the second exam is higher than the average score on the first exam in her statistics class. The output is in Figure 11.2.

SAS Program

```
filename exams 'grades.dat';

data grades;
   infile exams;
   input exam1 13-14 exam2 17-18;
   diff = exam2 - exam1;
run;
```

Figure 11.1 PROC TTEST Output for Example 11.1

```
              Comparing the Brake Response Time for            1
                  Cargo Vans in 40 mph Speed Zones
                              8:46 Thursday, May 20, 1993

                          TTEST PROCEDURE

Variable: RESPTIME        Response Time

GROUP        N           Mean            Std Dev         Std Error
-------------------------------------------------------------------
    1        73       41.17808219       15.62702650      1.82900511
    2        31       40.74193548       12.39077006      2.22544800

Variances        T        DF      Prob>|T|
-------------------------------------------
Unequal       0.1514     70.8      0.8801
Equal         0.1379    102.0      0.8906

For H0: Variances are equal, F' = 1.59    DF = (72,30)
                          Prob>F' = 0.1589




              Comparing the Brake Response Time for            2
                 Straight Trucks in 50 mph Speed Zones
                              8:46 Thursday, May 20, 1993

                          TTEST PROCEDURE

Variable: RESPTIME        Response Time

GROUP        N           Mean            Std Dev         Std Error
-------------------------------------------------------------------
    1        59       49.54237288       22.59868934      2.94209875
    2        32       65.59375000       28.69625060      5.07282835

Variances        T        DF      Prob>|T|
-------------------------------------------
Unequal      -2.7372     52.2      0.0085
Equal        -2.9371     89.0      0.0042

For H0: Variances are equal, F' = 1.61    DF = (31,58)
                          Prob>F' = 0.1160
```

```
proc means data=grades n mean std t prt;
   var diff;
title 'Is the Average Score on Exam 2 Higher than that on Exam 1?'
run;
```

New Statements

```
proc means data=grades n mean std t prt;
var diff;      The keywords T and PRT request a t-statistic and p-value for the
               variable DIFF.
```

Discussion of Output

The t-statistic is −0.03 with a p-value of 0.9762. Ms. Cohen would not reject the null hypothesis, and she would conclude that the average scores on the first and seconds exams are not different.

Figure 11.2 PROC MEANS Output for Example 11.2

```
Is the Average Score on Exam 2 Higher than that on Exam 1?   1
                                 11:24 Tuesday, April 12, 1994

     Analysis Variable : DIFF


   N         Mean        Std Dev           T  Prob>|T|
   -------------------------------------------------------
   49    -0.1632653     38.1320878    -0.0299710    0.9762
   -------------------------------------------------------
```

PROBLEMS

The files for these problems are described in detail in the Appendix.

11.1   Use the file cataract.dat to answer whether the average amount of astigmatism is different for folding and nonfolding lenses.

11.2   Use gas.dat to decide if the average gas mileage for manual transmissions is greater than the average for automatic transmissions.

11.3   Use the file grades.dat to test whether the average final exam grade is different for women and men.

11.4   Use the file handinj.dat for the following questions:

      a)   Is the average number of work days lost different for work and sports injuries?

b)   Is the average cost different for work and sports injuries?

11.5   Use the file src.dat to test whether women consider themselves to be more concerned about the environment than men as measured by the average environmentalist score.

11.6   Use the file robot.dat to test if humans and robots differ for

a)   throughput.
b)   quality.

YOU SHOULD NOW KNOW

how to use PROC TTEST to compare two means

how to use PROC MEANS to do a paired t-test

One-Way ANOVA
MODULES NEEDED 1, 2, 5

In analysis of variance or ANOVA, one is interested in testing whether a group of two or more means are equal. The variable that separates the data into these groups is referred to as a factor. One-way ANOVA tests whether the factor in a linear model is significant, that is, whether the means for the levels of that factor are equal.

The factor is an independent variable in the model, having two or more levels. The following is a model in one-way ANOVA with one factor (A) in the model:

$$Y_{ij} = \mu + A_i + \varepsilon_{ij}$$

PROC GLM and PROC ANOVA are two procedures for analyzing ANOVA models. In this module, you will learn how to use both procedures and the differences between them.

ANALYSIS OF VARIANCE—PROC GLM AND PROC ANOVA

Both PROC GLM and PROC ANOVA handle ANOVA problems. Their general forms are the same. They can be used interchangeably when the number of observations for each level of a factor are the same.

PROC GLM is an "all-purpose" procedure that can be used to analyze all types of general linear models (including regression and multivariate models). PROC GLM correctly handles unbalanced data in ANOVA. Data is "unbalanced" when the levels of the independent variables have unequal sample sizes. Because PROC GLM will do many things, it is generally not as efficient as PROC ANOVA when the data is balanced. This means that PROC GLM will use more computer resources to generate the same output as PROC ANOVA when the sample sizes are the same for each level of the independent variables. It is not appropriate to use PROC ANOVA when the data are unbalanced; it is appropriate to use PROC GLM when analyzing any kind of linear model.

The CLASS statement is used to designate which variables are factors in the model. The CLASS variables can have numeric values or character values. Without this statement, PROC GLM will assume that the independent variables designated in the model statement are numeric and will compute statistics for a regression model rather than an ANOVA model.

The MODEL statement has the form dependent variable = factor(s). The MODEL statement has many options, but none will be covered in this manual.

The MEANS statement requests multiple comparisons. List each class variable for which you want multiple comparisons after MEANS. More than one method of multiple comparisons can be listed, along with the option for grouping nonsignificant means together (LINES) and computing confidence intervals (CLDIFF).

PROC GLM General Form

```
proc glm data=data set;
   by variables;
   class variables;
   model dependent variable = independent variables;
   means effects </options>;
```

MEANS Option	Description
alpha=p	This is the significance level used in multiple comparisons. The default is .05.
bon	This requests Bonferroni t-tests of differences between means.
cldiff	This option requests confidence intervals for all pairwise differences between means.
duncan	This option requests Duncan's multiple comparisons.
lines	This option lists the means in descending order, indicating those means that are not significantly different with stars simulating a line segment beside them.
scheffe	This option requests Scheffé's multiple comparisons.
snk	This option requests the Student-Newman-Keuls multiple range test.
lsd	This option performs pairwise t-tests, which is equivalent to Fisher's least-significant-difference test when cell sizes are equal.
tukey	This option performs Tukey's studentized range test.

PROC ANOVA General Form

```
proc anova data=data set;
   by variables;
   class variables;
   model dependent variable = independent variables;
   means effects </options>;
```

Sums of Squares

SAS computes sums of squares four different ways. By default, SAS automatically prints Type I sums of squares and Type III sums of squares.

The Type I sums of squares are sequential sums of squares and are the ones described in most introductory statistics books. They have the property that the sums of squares from each component in the model add up to the total sum of squares. The hypothesis for each term in the model is a conditional test: It assumes any factors listed prior to the one being tested are already in the model.

The Type II sums of squares do not necessarily add up to the total sum of squares. Each hypothesis is conditioned on all the other terms being in the model. To get Type II sums of squares, use /SS2 in the MODEL statement.

The Type III sums of squares, printed by default with Type I sums of squares, are referred to as partial sums of squares. They do not normally add up to the total sum of squares. The hypotheses do not depend on the ordering of the effects in the model and do not involve parameters of other effects. Type III sums of squares are orthogonal if the design has missing cells.

The Type IV sums of squares are identical to the Type III sums of squares unless the design has missing cells. Then the Type IV sums of squares have a balancing property. To get Type IV sums of squares, use /SS4 in the MODEL statement.

EXAMPLE 12.1

Researchers wanted to test whether the average braking time of drivers following different types of trucks equipped with a center high-mounted stop lamp are the same or not. The output is shown in Figure 12.1.

SAS Program

```
filename inbrakes 'taillite.dat';

data one;
   infile inbrakes;
   input id vehtype group positn speedzn resptime
     follotme folltmec;

   if group = 1;

   label vehtype = 'Vehicle Type'
         group   = 'Group - Light On=1    Light Off=2'
         positn  = 'Light Position'
         speedzn = 'Speed Zone'
         resptime= 'Response Time'
         follotme= 'Following Time in Video Frames'
         folltmec= 'Following Time in Categories'
         ;
run;

proc glm data=one;
   class vehtype;
```

```
    model resptime = vehtype;
    means vehtype / tukey lines;
title 'One-Way ANOVA for Tail Light Study';
run;
```

New Statements

```
proc glm;
class vehtype;
```

> Because there are a different number of truck types, PROC GLM is the appropriate procedure to use here. The CLASS statement says that VEHTYPE is a factor or main effect in the ANOVA model.

```
model resptime = vehtype;
```

> The MODEL statement says there is only one factor in the model— VEHTYPE.

```
means vehtype / tukey lines;
```

> This MEANS statement requests Tukey multiple comparisons for the factor VEHTYPE. LINES groups nonsignificant means together.

Discussion of Output

Page 1 of the output (see Figure 12.1) says that there is one factor in the model— VEHTYPE—with four levels having values 1, 2, 3, and 4. There are 733 observations.

On page 2 of the output, the F-statistic for testing whether VEHTYPE is significant is 3.75. The p-value of the test is 0.0109. This p-value is shown in three places for one-way ANOVA: The first is for Model, and the other two are under Type I SS and Type III SS. Because the p-value is small, the null hypothesis of equal means for the different levels of VEHTYPE is rejected. The conclusion is that the average response times are different for different vehicle types.

Page 3 gives the Tukey multiple comparisons. There is a warning that the cell sizes are unequal and that the harmonic mean was used. The LINES statement produced the A and B groupings besides the means.

The conclusion for Example 12.1 is that the mean response times for the different truck types are significantly different. Straight trucks (4), minivans (3), and pickups (1) are not significantly different; minivans, pickups, and cargo vans (2) are not significantly different. Straight trucks and cargo vans are significantly different.

Figure 12.1 PROC GLM Output for Example 12.1

```
One-Way ANOVA for Tail Light Study              1
                        10:39 Monday, June 28, 1993

            General Linear Models Procedure
              Class Level Information

          Class    Levels    Values

          VEHTYPE     4      1 2 3 4


    Number of observations in data set = 733
```

```
            One-Way ANOVA for Tail Light Study          2
                        10:39 Monday, June 28, 1993

            General Linear Models Procedure

Dependent Variable: RESPTIME    Response Time
```

Source	DF	Sum of Squares	F Value	Pr > F
Model	3	3886.53768910	3.75	0.0109
Error	729	252025.52779521		
Corrected Total	732	255912.06548431		

R-Square	C.V.	RESPTIME Mean
0.015187	41.91847	44.35607094

```
            General Linear Models Procedure

Dependent Variable: RESPTIME    Response Time
```

Source	DF	Type I SS	F Value	Pr > F
VEHTYPE	3	3886.53768910	3.75	0.0109

Source	DF	Type III SS	F Value	Pr > F
VEHTYPE	3	3886.53768910	3.75	0.0109

Figure 12.1 cont. PROC GLM Output for Example 12.1

```
       One-Way ANOVA for Tail Light Study              3
                          10:39 Monday, June 28, 1993

            General Linear Models Procedure

Tukey's Studentized Range (HSD) Test for variable: RESPTIME

NOTE: This test controls the type I experimentwise error
      rate, but generally has a higher type II error rate
      than REGWQ.

           Alpha= 0.05  df= 729  MSE= 345.714
        Critical Value of Studentized Range= 3.642
          Minimum Significant Difference= 5.0402
             WARNING: Cell sizes are not equal.
          Harmonic Mean of cell sizes= 180.4668

Means with the same letter are not significantly different.


        Tukey Grouping            Mean      N  VEHTYPE

                     A           47.982    169  4
                     A
              B      A           45.053    189  3
              B      A
              B      A           42.962    157  1
              B
              B                  41.945    218  2
```

PROBLEMS

The files for these problems are described in detail in the Appendix.

12.1   Use the file taillite.dat to answer the following questions. If the ANOVA test is significant, do multiple comparisons.

      a)   For the experimental group in the 30 mph speed zone, are the average response times for truck types different?

      b)   For the experimental group and minivans, are the average response times for speed zones different?

12.2   Use the file wine.dat to answer the following questions. If the ANOVA test is significant, do multiple comparisons.

      a)   Do the average ratings differ for wine brand?

      b)   Do the average ratings differ for temperature?

12.3   Use the file calls.dat to answer the following questions. If the ANOVA test is significant, do multiple comparisons.

a)   Are the number of calls different for different shifts?
b)   Are the number of calls different for different days?

YOU SHOULD NOW KNOW

how to do one-way analysis of variance using PROC GLM and PROC ANOVA

Two-Way ANOVA
MODULES NEEDED 1, 2, 5, 12

In two-way ANOVA, there are two factors or main effects and possibly an interaction term in the model. A hypothesis test is associated with each term. The null hypothesis is that the main effect or interaction does not have an effect on the response variable. It states that the means for the different levels of the effect are the same.

PROC GLM and PROC ANOVA produce a table that contains the test statistics and p-values for all these hypotheses. In this module, you will learn how to modify the MODEL statement in PROC ANOVA and PROC GLM for two-way analysis of variance.

TWO-WAY ANOVA

In going from one-way ANOVA to two-way ANOVA, only the MODEL statement changes in both PROC GLM and PROC ANOVA. Both factors are listed along with any interaction term. The interaction term, or crossed effect, is designated with an asterisk (*) between the two main effects. You can also use a bar notation, which is shorthand for "do all crossing of factors." The following two MODEL statements reflect the SAS code for the model $Y_{ij} = \mu + A_i + B_j + (AB)_{ij}$:

```
model y = a b a*b;
model y = a | b;
```

The F-statistics assume that any factors in the model are fixed rather than random. It is also assumed that the model reflects a completely randomized design.

Other Designs

In PROC GLM, you can designate that some factors are random by using a RANDOM statement. This statement must follow the MODEL statement and has the form

```
random factors;
```

When this statement is used, the expected value of each of the different types of sums of squares is printed. This information is not used in the computation of the F-tests, however: They are still computed assuming that all factors are fixed. The sums of squares and means squares are computed correctly, and the expected means squares from the RANDOM statement provide the information to form the appropriate F-tests.

For a randomized block design, the blocking factor is not given any special designation. It appears in the model as a factor. It is up to the analyst to determine which of the printed statistics in the ANOVA table are appropriate to use.

Parentheses are used in the PROC GLM MODEL statement to designate nested effects. For example, if STATE and CITY are two factors in a model with CITY nested within STATE, the MODEL statement would be

```
model y = state city(state);
```

It is possible to perform a repeated measures design with PROC GLM. Because SAS handles this as a multivariate analysis (whether or not it really is) and because the data must be entered in a multivariate way, the discussion of how to do this analysis is beyond the scope of this introductory manual. The reader is referred to a SAS manual for further information.

EXAMPLE 13.1

In an experiment designed to compare the Hybrid III dummy to humans, two types of impactors (bar and disc) were used to fracture the skull. The response variable is stiffness (force vs. displacement in N/mm). Is the average stiffness of dummies and humans different? Is the average stiffness for type of impactor different? The output is shown in Figure 13.1.

SAS Program

```
filename stiff 'dummy.dat';

data force;
   infile stiff;
   input species $ impactor $ stiff1 stiff2 calcium magnesm;

   label species  = 'H=human and D=dummy'
         impactor = 'Type of Impactor'
         stiff1   = 'Stiffness Measure at Site 1'
         stiff2   = 'Stiffness Measure at Site 2'
         calcium  = '% Calcium in Bone'
         magnesm  = '% Magnesium in Bone'
         ;
run;

proc glm;
   class species impactor;
   model stiff1 = species impactor species*impactor;
   means species impactor / duncan lines;
```

```
title 'Two-Way ANOVA for Dummy Data';
run;
```

New Statements

```
class species impactor;
model stiff1 = species impactor species*impactor;
```
> The CLASS and MODEL statements indicate that there are two main effects (SPECIES and IMPACTOR) and an interaction in the model.

Discussion of Output

Page 1 of the printout (see Figure 13.1) shows that there are two class variables in the model—SPECIES with two levels and IMPACTOR with two levels.

Page 2 shows the ANOVA table. The interaction term (SPECIES*IMPACTOR) is not significant (p-value = 0.4358). At 0.05 significance, SPECIES is not significant, but IMPACTOR is (p-value = 0.0001). One would conclude that the average stiffness for humans and dummies does not differ but that the average stiffness for types of impactor does.

Page 3 does a Duncan's multiple comparison test on SPECIES (which is unnecessary because the F-test did not find that the two means were significantly different). Page 4 gives the Duncan's results for IMPACTOR, which agrees with the F-test that the two means are significantly different.

Figure 13.1 Output for Example 13.1

```
              Two-Way ANOVA for Dummy Data                    1

               General Linear Models Procedure
                    Class Level Information

                 Class     Levels    Values

                 SPECIES      2      d h

                 IMPACTOR     2      bar dsc


          Number of observations in data set = 24
```

Figure 13.1 cont.   Output for Example 13.1

```
                    Two-Way ANOVA for Dummy Data                    2

                    General Linear Models Procedure

Dependent Variable: STIFF1    Stiffness Measure at Site 1

Source                   DF      Sum of Squares   F Value    Pr > F

Model                     3      26643390.4583     17.38     0.0001

Error                    20      10222149.1667

Corrected Total          23      36865539.6250

                  R-Square              C.V.         STIFF1 Mean

                  0.722718           23.40444      3054.6250000


Source                   DF        Type I SS   F Value    Pr > F

SPECIES                   1      1839834.3750      3.60    0.0723
IMPACTOR                  1     24480380.0417     47.90    0.0001
SPECIES*IMPACTOR          1       323176.0417      0.63    0.4358

Source                   DF       Type III SS  F Value    Pr > F

SPECIES                   1      1839834.3750      3.60    0.0723
IMPACTOR                  1     24480380.0417     47.90    0.0001
SPECIES*IMPACTOR          1       323176.0417      0.63    0.4358

                    Two-Way ANOVA for Dummy Data                    3
                              14:33 Monday, June 28, 1993


                    General Linear Models Procedure
              Duncan's Multiple Range Test for variable: STIFF1

     NOTE: This test controls the type I comparisonwise error
           rate, not the experimentwise error rate

              Alpha= 0.05  df= 20  MSE= 511107.5

                    Number of Means      2
                    Critical Range   608.0

     Means with the same letter are not significantly different.

          Duncan Grouping            Mean     N  SPECIES

                           A        3331.5     12  h
                           A
                           A        2777.8     12  d
```

Figure 13.1 cont. Output for Example 13.1

```
                Two-Way ANOVA for Dummy Data              4

                General Linear Models Procedure

       Duncan's Multiple Range Test for variable: STIFF1

    NOTE: This test controls the type I comparisonwise error
          rate, not the experimentwise error rate

              Alpha= 0.05  df= 20  MSE= 511107.5

                   Number of Means     2
                   Critical Range   608.0

   Means with the same letter are not significantly different.

       Duncan Grouping            Mean     N  IMPACTOR

              A                  4064.6    12  bar

              B                  2044.7    12  dsc
```

PROBLEMS

The files for these problems are described in detail in the Appendix.

13.1   Use the file taillite.dat to answer the following questions:

a)   Do a two-way ANOVA, including the interaction term with response time as the dependent variable and group and vehicle type as the class variables.

b)   Do a two-way ANOVA, including the interaction term with response time as the dependent variable and group and speed zone as the class variables.

c)   Do a two-way ANOVA, including the interaction term with response time as the dependent variable and group and CHMSL position as the class variables.

13.2   Use the file brownie.dat to see if there is a difference in the amount of hard crust for type of pan and mix brand. Include an interaction term. What are your conclusions?

13.3   Use the file wear.dat to see if wear depends on cut depth or speed. Include an interaction term. What are your conclusions?

13.4   Use the file calls.dat to test whether shift, day, or the interaction are significant on the number of calls received. What are your conclusions?

YOU SHOULD NOW KNOW

how to use PROC GLM and PROC ANOVA to do a two-way analysis of variance

Correlations
MODULES NEEDED 1, 2, 5

Whenever you have bivariate random variables, you may be interested in how much they are related. You may want to compute correlation coefficients or partial correlation coefficients. In this module, you will learn how to use PROC CORR.

PROC CORR

This procedure computes the Pearson correlation coefficient and produces simple descriptive statistics—mean, standard deviation, sum, minimum, and maximum—for all pairs of variables listed in the VAR statement. It also computes a p-value for testing whether the true correlation $\rho=0$. The correlations are given in matrix form.

    The WITH statement is used to reduce the number of correlation coefficients that are computed. The pairs of variables come from those listed in the VAR statement crossed with those listed in the WITH statement. The total number of pairs would be the number of variables in the VAR statement times the number in the WITH statement.

    The PARTIAL statement produces partial correlation coefficients for the variables listed in the VAR statement, controlling for the effect of the variables listed in the PARTIAL statement.

PROC CORR General Form

```
proc corr data=data set options;
   by variables;
   var variables;
   with variables;
   partial variables;
```

PROC CORR Option	Description
cov	Prints the covariance matrix.
nosimple	Suppresses the printing of simple descriptive statistics.
noprob	Suppresses the printing of p-values for H_o: $\rho=0$ vs. H_a: $\rho\neq0$.

EXAMPLE 14.1

Renee Smith developed an alternative method of assessing the quality of bones used in transplants. She wanted to know how well her method correlated with the standard method, measured by a % young normal score.

```
filename scores 'bonescor.dat';

data one;
   infile scores;
   input singh ccratio csi calcar bonescor young;

   label singh    = 'Singh index'
         ccratio  = 'CC ratio'
         csi      = 'CSI'
         calcar   = 'Calcar width'
         bonescor = 'Renee''s bone score'
         young    = '% Young Normal'
         ;
run;

proc corr data=one;
   var bonescor young singh ccratio csi calcar;
title 'Correlation of Renee''s bone score and % young normal';
run;
```

New Statements

```
proc corr data=one;
var bonescor young singh ccratio csi calcar;
```
There are 6 variables in this VAR statement, resulting in 15 pairs of variables and 15 correlation coefficients.

Discussion of Output

In Figure 14.1, the first page of the output gives the simple descriptive statistics for all six variables. On the second page of the output, you see that sample correlation coefficient for YOUNG and SINGH is 0.18908, and the p-value for testing whether the true correlation is significantly different from 0 is 0.3170. The correlation coefficient for BONESCOR and YOUNG is 0.29393 with a p-value of 0.1149.

Figure 14.1 Output from PROC CORR of Example 14.1

```
            Correlation of Renee's bone score and % young normal     1
                              10:06 Wednesday, June 30, 1993


                         CORRELATION ANALYSIS

    6 'VAR' Variables:    BONESCOR YOUNG     SINGH     CCRATIO
                          CSI      CALCAR

                    Simple Statistics

    Variable          N         Mean       Std Dev         Sum

    BONESCOR         30       7.7000        1.9325       231.0
    YOUNG            30       112.8        22.4420      3383.9
    SINGH            30       4.5000        1.0422       135.0
    CCRATIO          30       0.6070        0.1181      18.2100
    CSI              30       0.5967        0.0718      17.9000
    CALCAR           30       6.7000        1.6846       201.0


                    Simple Statistics

    Variable     Minimum     Maximum     Label

    BONESCOR      3.0000     10.0000     Renee's bone score
    YOUNG        73.9000      155.8      % Young Normal
    SINGH         2.0000      6.0000     Singh index
    CCRATIO       0.4300      0.9400     CC ratio
    CSI           0.4600      0.7400     CSI
    CALCAR        4.0000     10.0000     Calcar width
```

Figure 14.1 cont.   Output from PROC CORR of Example 14.1

```
      Correlation of Renee's bone score and % young normal      2
                           10:06 Wednesday, June 30, 1993


                      CORRELATION ANALYSIS


 Pearson Correlation Coefficients / Prob > |R| under Ho: Rho=0
 / N = 30
```

	BONESCOR	YOUNG	SINGH
BONESCOR	1.00000	0.29393	0.36810
Renee's bone score	0.0	0.1149	0.0453
YOUNG	0.29393	1.00000	0.18908
% Young Normal	0.1149	0.0	0.3170
SINGH	0.36810	0.18908	1.00000
Singh index	0.0453	0.3170	0.0
CCRATIO	-0.64490	-0.33175	0.22840
CC ratio	0.0001	0.0733	0.2248
CSI	0.77796	0.53198	0.10139
CSI	0.0001	0.0025	0.5940
CALCAR	0.58575	0.54714	0.16694
Calcar width	0.0007	0.0018	0.3779

	CCRATIO	CSI	CALCAR
BONESCOR	-0.64490	0.77796	0.58575
Renee's bone score	0.0001	0.0001	0.0007
YOUNG	-0.33175	0.53198	0.54714
% Young Normal	0.0733	0.0025	0.0018
SINGH	0.22840	0.10139	0.16694
Singh index	0.2248	0.5940	0.3779
CCRATIO	1.00000	-0.64605	-0.44852
CC ratio	0.0	0.0001	0.0129
CSI	-0.64605	1.00000	0.59020
CSI	0.0001	0.0	0.0006
CALCAR	-0.44852	0.59020	1.00000
Calcar width	0.0129	0.0006	0.0

PROBLEMS

The files for these problems are described in detail in the Appendix.

14.1   Use the file electric.dat to compute correlations for

a)   house size and appliance index. Is $\rho=0$?
b)   number in family and applicance index. Is $\rho=0$?
c)   house size and income. Is $\rho=0$?

14.2   Use the file gas.dat to compute correlations for

a)   gas mileage, horsepower, and torque. How is mileage related to horsepower and torque?
b)   car weight, car length, and car width. Describe the relationship between these three variables.

14.3   Use the file grades.dat to compute correlations for

a)   first exam, second exam, computer grade, quiz, and final exam grade. One variable is not strongly correlated with the others. Which one is it?
b)   first exam, second exam, final exam with quiz and computer grade.

14.4   Use the file handinj.dat to compute the correlation for days of work lost and cost.

14.5   Use the file utility.dat to compute correlations for phone costs, fuel costs, and electric costs. Which pair of variables has the strongest correlation?

YOU SHOULD NOW KNOW

how to use PROC CORR to compute correlation coefficients

Simple Linear Regression

MODULES NEEDED   1, 2, 5, 9

Regression is the area of statistics that is concerned with finding a model that describes the relationship between a dependent variable and one or more independent variables. The regression procedure in SAS is called PROC REG. PROC GLM will also do regression analysis, but PROC REG has more options that are useful for a regression analysis. PROC REG has many optional statements, some of which will be addressed in this and the next three modules.

In this module, you will learn how to create a scattergram, or plot, of the dependent variable (on the Y-axis) and one independent variable (on the X-axis); how to compute the least-squares regression line and output predicted values; and then plot the predicted values on the same plot as the original scattergram. You will also learn how to compute confidence intervals for Y and prediction intervals for future values of Y.

PROC REG

PROC REG uses a MODEL statement to define the theoretical model for the relationship between the independent and dependent variables. It is possible to have more than one MODEL statement in the PROC step. This is useful if you are trying to find which model best fits the data.

Using the PLOT statement, it is possible to create a plot that shows the predicted regression line. It works very much like PROC PLOT (see Module 9). The variables that you can use in the PLOT statement are those listed in the MODEL statement and some special variables created by SAS. Some of these are P. (the predicted values), R. (the residuals), STUDENT. (the studentized residuals), L95. and U95. (the lower and upper limits for prediction intervals), and L95M. and U95M. (the lower and upper limits for confidence intervals on the mean). Note that the dot (.) is part of the variable names: P., R., STUDENT., and so on.

The /OVERLAY option tells SAS to print any requested plots as one plot. Without /OVERLAY, separate plots would be generated. Any character, number, or symbol on the keyboard can be used as the plotting character. SAS may not be able to plot all the points in the data set because of the scale it chooses to fit the plot on one page. Sometimes two points may overlap or almost overlap. If this happens, SAS will choose another character to indicate that more than one point is plotted.

PROC REG General Form

```
proc reg data=data set <options>;
   by variables;
   model dependent variable=independent variables </options>;
   plot yvariable*xvariable <=symbol> </options>;
```

PROC REG Option	Description
simple	This option prints descriptive statistics for each variable in the MODEL statement: sum, mean, variance, and standard deviation.
noprint	This option suppresses the printed output. It is useful when creating an output data set. See Module 16.

MODEL Option	Description
p	This option will print the observed value of the dependent variable, the predicted value, and the residual for each observation in the data set.
r	This option prints everything the P-option prints plus the standard errors of the predicted values and residuals, the studentized residuals, and Cook's D-statistic. It also creates a small residual plot vs. the observation number where asterisks form a "bar" that is the length of the value of the residual.
clm	This option prints 95% confidence intervals for the mean of each observation.
cli	This option prints 95% prediction intervals for each observation.
selection=forward	This option performs forward regression. See Module 18.
selection=backward	This option performs backward regression. See Module 18.
selection=stepwise	This option performs stepwise regression. See Module 18.
selection=rsquare	This option will compute R^2, adjusted R^2, Mallow's C_p, and MSE for all possible models when the corresponding keywords are used. See Module 18.
sle=	This is the significance level for a variable to enter the model during forward or stepwise regression. The default is .5 for forward regression and .15 for stepwise regression. See Module 18.
sls=	This is the significance level for a variable to stay in the model during backward or stepwise regression. The default is .10 for backward regression and .15 for stepwise regression. See Module 18.
cp	This asks for Mallow's C_p statistic and is available only with SELECTION=RSQUARE. See Module 18.

adjrsq	This asks for the adjusted R^2 and is available only with SELECTION=RSQUARE. See Module 18.
mse	This asks for values of the mean square error and is available only with SELECTION=RSQUARE. See Module 18.

EXAMPLE 15.1

It is suggested that the ln of concentration of DNase, a recombinant protein in rat serum, can predict optical density, a measure of absorbance. To compute a least-squares regression line and see graphically how well it fits the data, use the following program. The output is shown in Figure 15.1.

SAS Program

```
filename rats 'assay.dat';

data protein;
   infile rats;
   input concen absorb;
   concen = log(concen);

   label concen   = 'Ln of concentration of DNase'
         absorb   = 'Optical Density'
         ;
run;

proc reg data=protein;
   model absorb = concen / p;
   plot absorb*concen='+' p.*concen='*' / overlay;
title 'Regression Line and Scatterplot for DNase Data';
run;
```

New Statements

```
proc reg;
model absorb = concen / p;
```
This is the MODEL statement for the relationship ABSORB $= \beta_0 + \beta_1$*CONCEN $+ \varepsilon$. Because of the /P option, the output will include observed, predicted, and residual values for each observation.

```
plot absorb*concen='+' p.*concen='*' / overlay;
```
This PLOT statement will produce a scatterplot of ABSORB vs. CONCEN with '+' as the plotting symbol. The estimated regression line (P.*CONCEN) is plotted using '*' as the plotting symbol. The option /OVERLAY tells SAS to print both of these plots as one plot.

Discussion of Output

Look at the first page of the output (Figure 15.1). The ANOVA table reports the F-statistic for testing $Ho:\beta_1=0$. Here the value is 822.037 with a p-value = 0.0001. The first page also has R^2 and adjusted R^2 values and t-tests for testing whether $\beta_1=0$ versus $\beta_1\neq0$, along with associated p-values. The estimated regression line is ABSORB = 0.6342 + 0.4135*CONCEN.

On the second page of the output are the results of using the /P option in the MODEL statement: Each value of the dependent variable is printed, along with the predicted value and the residual. You may want to consider the number of observations in your data set before using the /P option. At the bottom of the page is the sum of the residuals, which is 0, as it should be. The Press statistic is used to evaluate how "good" the regression model is as a predictive model.

The third page gives the scatterplot and regression line. Notice the question mark (?). This indicates that two or more points overlap. In each case, the question mark includes the predicted value because an asterisk (*) is not visible. Using a straight edge, you can connect the asterisks to see the estimated regression line.

Figure 15.1 PROC REG Output for Example 15.1

```
                 Regression Line and Scatterplot for DNase Data          1
                                            8:34 Tuesday, June 29, 1993
Model: MODEL1
Dependent Variable: ABSORB       Optical Density

                         Analysis of Variance

                          Sum of          Mean
    Source        DF      Squares        Square      F Value      Prob>F

    Model          1     13.80071      13.80071      822.037      0.0001
    Error         40      0.67154       0.01679
    C Total       41     14.47224

          Root MSE        0.12957     R-square       0.9536
          Dep Mean        0.81874     Adj R-sq       0.9524
          C.V.           15.82559

                          Parameter Estimates

                     Parameter      Standard     T for H0:
    Variable  DF     Estimate         Error     Parameter=0     Prob > |T|

    INTERCEP   1     0.634200      0.02100359     30.195          0.0001
    CONCEN     1     0.413495      0.01442198     28.671          0.0001

                        Variable
    Variable  DF       Label
    INTERCEP   1    Intercept
    CONCEN     1    Ln of concentration of DNase
```

Figure 15.1 cont. PROC REG Output for Example 15.1

Regression Line and Scatterplot for DNase Data   2
                                              8:34 Tuesday, June 29, 1993

Obs	Dep Var ABSORB	Predict Value	Residual
1	0.1590	-0.0411	0.2001
2	0.1550	-0.0411	0.1961
3	0.1370	-0.0411	0.1781
4	0.1230	-0.0411	0.1641
5	0.1520	-0.0411	0.1931
6	0.1480	-0.0411	0.1891
7	0.2460	0.2455	0.000488
8	0.2520	0.2455	0.00649
9	0.2250	0.2455	-0.0205
10	0.2070	0.2455	-0.0385
11	0.2260	0.2455	-0.0195
12	0.2220	0.2455	-0.0235
13	0.4270	0.5321	-0.1051
14	0.4110	0.5321	-0.1211
15	0.4010	0.5321	-0.1311
16	0.3830	0.5321	-0.1491
17	0.3920	0.5321	-0.1401
18	0.3830	0.5321	-0.1491
19	0.7040	0.8187	-0.1147
20	0.6840	0.8187	-0.1347
21	0.6720	0.8187	-0.1467
22	0.6810	0.8187	-0.1377
23	0.6580	0.8187	-0.1607
24	0.6440	0.8187	-0.1747
25	0.9940	1.1054	-0.1114
26	0.9800	1.1054	-0.1254
27	1.1160	1.1054	0.0106
28	1.0780	1.1054	-0.0274
29	1.0430	1.1054	-0.0624
30	1.0020	1.1054	-0.1034
31	1.4210	1.3920	0.0290
32	1.3850	1.3920	-0.00696
33	1.5540	1.3920	0.1620
34	1.5260	1.3920	0.1340
35	1.4660	1.3920	0.0740
36	1.3810	1.3920	-0.0110
37	1.7150	1.6786	0.0364
38	1.7210	1.6786	0.0424
39	1.9320	1.6786	0.2534
40	1.9140	1.6786	0.2354
41	1.7430	1.6786	0.0644
42	1.7240	1.6786	0.0454

Sum of Residuals   5.77316E-15
Sum of Squared Residuals   0.6715
Predicted Resid SS (Press)   0.7520

Figure 15.1 cont.   PROC REG Output for Example 15.1

```
Regression Line and Scatterplot for DNase Data               3
                                              8:34 Tuesday, June 29, 1993

      -+------+------+------+------+------+------+------+------+------+-
 ABSORB |                                                            |
      |                                                              |
  2.0 +                                                            + |
      |                                                           +|
      |                                                            |
      |                                                          ?|
      |                                                            |
  1.5 +                                              +           + |
      |                                              +             |
      |                                              +             |
      |                                              ?             |
  O   |                                                            |
  p   |                                                            |
  t   |                                   +                        |
  i   |                                   ?                        |
  c 1.0 +                                 +                      + |
  a   |                                                            |
  l   |                                                            |
      |                              *                             |
  D   |                              +                             |
  e   |                              +                             |
  n   |                                                            |
  s 0.5 +                    *                                   + |
  i   |                       +                                    |
  t   |                       +                                    |
  y   |               +                                            |
      |               ?                                            |
      |         +                                                  |
  0.0 +         *                                                + |
      |                                                            |
      |                                                            |
      |                                                            |
      |                                                            |
 -0.5 +                                                          + |
      |                                                            |
      |                                                            |
      -+------+------+------+------+------+------+------+------+------+-
      -2.0   -1.5   -1.0   -0.5    0.0    0.5    1.0    1.5    2.0   2.5

              Ln of concentration of DNase    CONCEN
```

EXAMPLE 15.2

Ms. Smith would now like to generate confidence intervals for the mean of each value of
ABSORB and prediction intervals for future values of ABSORB. The first page of the
output is the same as in Example 15.1. For the new information, see Figure 15.2.

SAS Program

```
filename rats 'assay.dat';

data protein;
   infile rats;
   input concen absorb;
   concen = log(concen);

   label concen  = 'Ln of concentration of DNase'
         absorb  = 'Optical Density'
         ;
run;

proc reg data=protein;
   model absorb = concen / clm cli;
title 'Confidence Intervals and Prediction Intervals';
title2 'for DNase Data';
run;
```

Discussion of Output

For each of the 42 data points, Figure 15.2 provides the observed value (Dep Var), the
predicted value (Predict Value), the standard error of the predicted value (Std Err Predict),
95% confidence interval limits for the true mean value (Lower95% Mean and Upper95%
Mean), 95% prediction interval limits for some future value (Lower95% Predict and
Upper95% Predict), and the residual.

Figure 15.2 PROC REG Output for Example 15.2

Confidence Intervals and Prediction Intervals   2
for DNase Data   8:34 Tuesday, June 29, 1993

Obs	Dep Var ABSORB	Predict Value	Std Err Predict	Lower95% Mean	Upper95% Mean	Lower95% Predict	Upper95% Predict
1	0.1590	-0.0411	0.036	-0.1139	0.0317	-0.3129	0.2307
2	0.1550	-0.0411	0.036	-0.1139	0.0317	-0.3129	0.2307
3	0.1370	-0.0411	0.036	-0.1139	0.0317	-0.3129	0.2307
4	0.1230	-0.0411	0.036	-0.1139	0.0317	-0.3129	0.2307
5	0.1520	-0.0411	0.036	-0.1139	0.0317	-0.3129	0.2307
6	0.1480	-0.0411	0.036	-0.1139	0.0317	-0.3129	0.2307
7	0.2460	0.2455	0.028	0.1884	0.3027	-0.0225	0.5135
8	0.2520	0.2455	0.028	0.1884	0.3027	-0.0225	0.5135
9	0.2250	0.2455	0.028	0.1884	0.3027	-0.0225	0.5135
10	0.2070	0.2455	0.028	0.1884	0.3027	-0.0225	0.5135
11	0.2260	0.2455	0.028	0.1884	0.3027	-0.0225	0.5135
12	0.2220	0.2455	0.028	0.1884	0.3027	-0.0225	0.5135
13	0.4270	0.5321	0.022	0.4869	0.5773	0.2664	0.7979
14	0.4110	0.5321	0.022	0.4869	0.5773	0.2664	0.7979
15	0.4010	0.5321	0.022	0.4869	0.5773	0.2664	0.7979
16	0.3830	0.5321	0.022	0.4869	0.5773	0.2664	0.7979
17	0.3920	0.5321	0.022	0.4869	0.5773	0.2664	0.7979
18	0.3830	0.5321	0.022	0.4869	0.5773	0.2664	0.7979
19	0.7040	0.8187	0.020	0.7783	0.8591	0.5538	1.0837
20	0.6840	0.8187	0.020	0.7783	0.8591	0.5538	1.0837
21	0.6720	0.8187	0.020	0.7783	0.8591	0.5538	1.0837
22	0.6810	0.8187	0.020	0.7783	0.8591	0.5538	1.0837
23	0.6580	0.8187	0.020	0.7783	0.8591	0.5538	1.0837
24	0.6440	0.8187	0.020	0.7783	0.8591	0.5538	1.0837
25	0.9940	1.1054	0.022	1.0602	1.1505	0.8396	1.3711
26	0.9800	1.1054	0.022	1.0602	1.1505	0.8396	1.3711
27	1.1160	1.1054	0.022	1.0602	1.1505	0.8396	1.3711
28	1.0780	1.1054	0.022	1.0602	1.1505	0.8396	1.3711
29	1.0430	1.1054	0.022	1.0602	1.1505	0.8396	1.3711
30	1.0020	1.1054	0.022	1.0602	1.1505	0.8396	1.3711
31	1.4210	1.3920	0.028	1.3348	1.4491	1.1239	1.6600
32	1.3850	1.3920	0.028	1.3348	1.4491	1.1239	1.6600
33	1.5540	1.3920	0.028	1.3348	1.4491	1.1239	1.6600
34	1.5260	1.3920	0.028	1.3348	1.4491	1.1239	1.6600
35	1.4660	1.3920	0.028	1.3348	1.4491	1.1239	1.6600
36	1.3810	1.3920	0.028	1.3348	1.4491	1.1239	1.6600
37	1.7150	1.6786	0.036	1.6057	1.7514	1.4068	1.9504

Figure 15.2 cont.   PROC REG Output for Example 15.2

```
          Confidence Intervals and Prediction Intervals              3
                      for DNase Data      8:34 Tuesday, June 29, 1993
```

Obs	Dep Var ABSORB	Predict Value	Std Err Predict	Lower95% Mean	Upper95% Mean	Lower95% Predict	Upper95% Predict
38	1.7210	1.6786	0.036	1.6057	1.7514	1.4068	1.9504
39	1.9320	1.6786	0.036	1.6057	1.7514	1.4068	1.9504
40	1.9140	1.6786	0.036	1.6057	1.7514	1.4068	1.9504
41	1.7430	1.6786	0.036	1.6057	1.7514	1.4068	1.9504
42	1.7240	1.6786	0.036	1.6057	1.7514	1.4068	1.9504

Obs	Residual
1	0.2001
2	0.1961
3	0.1781
4	0.1641
5	0.1931
6	0.1891
7	0.000488
8	0.00649
9	-0.0205
10	-0.0385
11	-0.0195
12	-0.0235
13	-0.1051
14	-0.1211
15	-0.1311
16	-0.1491
17	-0.1401
18	-0.1491
19	-0.1147
20	-0.1347
21	-0.1467
22	-0.1377
23	-0.1607
24	-0.1747
25	-0.1114
26	-0.1254
27	0.0106
28	-0.0274

Figure 15.2 cont.   PROC REG Output for Example 15.2

```
              Confidence Intervals and Prediction Intervals         4
                      for DNase Data    8:34 Tuesday, June 29, 1993


    Obs   Residual

     29    -0.0624
     30    -0.1034
     31     0.0290
     32    -0.00696
     33     0.1620
     34     0.1340
     35     0.0740
     36    -0.0110
     37     0.0364
     38     0.0424
     39     0.2534
     40     0.2354
     41     0.0644
     42     0.0454


Sum of Residuals              5.77316E-15
Sum of Squared Residuals         0.6715
Predicted Resid SS (Press)       0.7520
```

PROBLEMS

The files for these problems are described in detail in the Appendix.

15.1   Use the file bonscor.dat using bone score as the dependent variable and % young normal as the independent variable to

   a)   compute the estimated regression line.
   b)   plot the data and overlay the estimated regression line.
   c)   comment on how well the line fits the data.

15.2   Use the file electric.dat to compute and plot estimated regression lines using peak load as the dependent variable for the following models. Comment on how well the line fits the data.

   a)   Let air capacity be the independent variable.
   b)   Use appliance index as the independent variable.
   c)   Use the number of family members as the independent variable.

15.3   Use the file gas.dat to compute and plot the estimated regression line for the following models. Comment on how well the line fits the data.

   a)   Dependent variable is power and independent variable is displacement.
   b)   Dependent variable is torque and independent variable is displacement.
   c)   Dependent variable is torque and independent variable is power.
   d)   Dependent variable is mileage and independent variable is displacement.
   e)   Dependent variable is mileage and independent variable is torque.
   f)   Dependent variable is mileage and independent variable is power.

15.4   a)   Compute confidence intervals for the average Y for Problem 15.2(a).
        b)   Compute prediction intervals for future values of Y for Problem 15.2(b).

15.5   Compute the following intervals:

   a)   Confidence intervals for the average Y for Problem 15.3(a).
   b)   Prediction intervals for future values of Y for Problem 15.3(b).
   c)   Confidence intervals for the average Y for Problem 15.3(c).
   d)   Prediction intervals for future values of Y for Problem 15.3(d).
   e)   Confidence intervals for the average Y for Problem 15.3(e).
   f)   Prediction intervals for future values of Y for Problem 15.3(f).

YOU SHOULD NOW KNOW

how to find the estimated regression line

how to plot the estimated regression line with the data

how to generate confidence intervals and prediction intervals

Model Checking in Regression
MODULES NEEDED   1, 2, 5, 9, 15

In regression analysis, assumptions are made about the error terms. Specifically, it is assumed that errors are independent and normally distributed with mean 0 and variance σ^2. Once a least-squares regression line is computed, it is necessary to check the model assumptions by analyzing the residuals and looking at some special plots of the data.

In this module, you will learn how to produce these plots.

PLOTS FOR MODEL CHECKING

Several plots can be created that are useful in verifying model assumptions. They are

1. y vs. y-hat. Since y-hat is used to predict y for each x value, this plot should look like a 45° line when the X and Y axes are scaled the same.

2. residual e (or standardized residual) vs. y-hat. Since e is an estimate of ε for each x, this plot should show a random scatter about 0. The width of the scatter should be the same if the assumption of constant variance is true.

3. residual e (or standardized residual) vs. x. Since e is an estimate of ε for each x, this plot should show a random scatter about 0. The width of the scatter should be the same if the assumption of constant variance is true. This plot should be done for each independent variable in the model.

4. normal probability plot of errors. This will check the normality assumption.

MORE ON PROC REG

Many SAS PROCs have the ability to create a new SAS data set that contains information generated by the PROC. PROC REG can create a data set containing the original data of x's and y's, plus the predicted values and standardized residuals if requested by the appropriate keywords. This data set can also contain the upper and lower bounds for confidence intervals and prediction intervals. The statement that produces a new data set is OUTPUT OUT.

PROC REG's OUTPUT OUT Statement General Form

```
output out=new data set keyword=names <...keyword=names>;
```

Possible Keyword	Description
p=*varname*	The predicted values will be included in the output data set, where VARNAME is the name you want assigned to the predicted values.
r=*varname*	The residuals will be included in the output data set, where VARNAME is the name you want assigned to the residuals.
student=*varname*	Standardized residuals (also called studentized residuals) will be included in the output data set, where VARNAME is the name you assign to the standardized residuals.
L95M=*varname*	The lower bound for 95% confidence intervals is included in the output data set.
U95M=*varname*	The upper bound for 95% confidence intervals is included in the output data set.
L95=*varname*	The lower bound for 95% prediction intervals is included in the output data set.
U95=*varname*	The upper bound for 95% prediction intervals is included in the output data set.

EXAMPLE 16.1

Here's how to check the model from Example 15.1. The output is shown in Figure 16.1.

SAS Code

```
filename rats 'assay.dat';

data protein;
   infile rats;
   input concen absorb;
   concen = log(concen);

   label concen    = 'Ln of concentration of DNase'
         absorb    = 'Optical Density'
         ;
run;

proc reg data=protein noprint;
   model absorb = concen;
   output out=new p=yhat student=resid;
run;
```

```
proc plot data=new;
   plot absorb*yhat;
   title 'Model Checking - Y vs. Y-hat';

   plot resid*yhat;
   title 'Model Checking - Standardized Residuals vs. Y-hat';

   plot resid*concen;
   title 'Model Checking - Standardized Residuals vs. X';
run;

proc univariate normal plot data=new;
   var resid;
title 'Model Checking - Normal Test and Plot';
run;
```

New Statements

```
output out=new p=yhat student=resid;
```
        The OUTPUT OUT= statement creates a new data set named NEW containing the x's and y's and the predicted values and standardized residuals, which were requested with the keywords P= and R=, respectively.

Discussion of Output

In looking at the output in Figure 16.1, the plot of y vs. y-hat does not show a straight line; instead, it looks curvilinear. For the second and third plots, we do not see random scatter about zero: The pattern shows residuals that are positive, then negative, and then positive.

These plots indicate that the linear model we chose is not a good estimate of the true relationship between the dependent variables ln of concentration of DNase and the absorbance measure optical density.

Not shown is the printed output from PROC UNIVARIATE (discussed in Module 5). The p-value for testing whether the data is normal is 0.0019, indicating nonnormality.

Figure 16.1 Output for Model Checking from Example 16.1

```
            Model Checking - Standardized Residuals vs. X              1
                                        8:34 Tuesday, June 29, 1993

        Plot of ABSORB*YHAT.  Legend: A = 1 obs, B = 2 obs, etc.

       |
       |
   2.0 +
       |                                                           B
       |
       |                                                           D
       |
 O     |                                               A
 p 1.5 +                                               B
 t     |                                               A
 i     |                                               B
 c     |
 a     |
 l     |                                      A
       |                                      B
 D 1.0 +                                      C
 e     |
 n     |
 s     |
 i     |                            C
 t     |                            C
 y     |
   0.5 +
       |                     C
       |                     C
       |           A
       |           E
       |F
   0.0 +
       |
        -+----------+----------+----------+----------+----------+----------+-
       -0.041     0.246      0.532      0.819      1.105      1.392      1.679

                           Predicted Value of ABSORB
```

Figure 16.1 cont. Output for Model Checking from Example 16.1

```
            Model Checking - Standardized Residuals vs. X          2
                                        8:34 Tuesday, June 29, 1993

         Plot of RESID*YHAT.  Legend: A = 1 obs, B = 2 obs, etc.

        |
        |
    2.0 +                                                         A
        |                                                         A
        |
        |C
    1.5 +A
        |A
S       |A                                          A
t       |
u   1.0 +                                           A
d       |
e       |
n       |
t       |                                           A
i       |
z   0.5 +                                                         A
e       |                                                         A
d       |                                                         B
        |                                           A
R       |               A                A
e   0.0 +               A
s       |                                             B
i       |               C                A
d       |               A
u       |
a  -0.5 +                                A
l       |
        |
        |            A                   A
        |            A          A        A
   -1.0 +            A                   A
        |            A          C
        |            B
        |                       A
        |                       A
   -1.5 +
        |
        -+----------+----------+----------+----------+----------+----------+
       -0.041     0.246      0.532      0.819      1.105      1.392      1.679

                        Predicted Value of ABSORB
```

Figure 16.1 cont. Output for Model Checking from Example 16.1

```
                    Model Checking - Standardized Residuals vs. X              3
                                                  8:34 Tuesday, June 29, 1993


                  Plot of RESID*CONCEN.   Legend: A = 1 obs, B = 2 obs, etc.

         |
         |
    2.0 +|                                                                    A
         |                                                                    A
         |
         |C
    1.5 +A
         |A
         |A                                          A
S        |
t   1.0 +|                                           A
u        |
d        |
e        |
n        |                                           A
t   0.5 +|                                                                    A
i        |                                                                    A
z        |                                                                    B
e        |
d        |                                           A
    0.0 +|          A                       A
R        |          A
e        |                                           B
s        |          C                       A
i        |          A
d  -0.5 +|                                   A
u        |
a        |
l        |              A                   A
         |              A        A          A
   -1.0 +|              A                   A
         |              A        C
         |              B
         |                       A
         |                       A
   -1.5 +|
         |
        -+----------+----------+----------+----------+----------+----------+
       -1.633    -0.940     -0.247     0.446      1.139      1.833     2.526

                            Ln of concentration of DNase
```

PROBLEMS

The files for these problems are described in detail in the Appendix.

16.1   a)   Check the model assumptions for Problem 15.2(a).
       b)   Check the model assumptions for Problem 15.2(b).
       c)   Check the model assumptions for Problem 15.2(c).

16.2   a)   Check the model assumptions for Problem 15.3(a).
       b)   Check the model assumptions for Problem 15.3(b).
       c)   Check the model assumptions for Problem 15.3(c).
       d)   Check the model assumptions for Problem 15.3(d).
       e)   Check the model assumptions for Problem 15.3(e).
       f)   Check the model assumptions for Problem 15.3(f).

16.3   Use the file grades.dat to run regression models and check assumptions, using final exam as the dependent variable, and for the independent variable use

       a)   first exam.
       b)   second exam.
       c)   quiz.

YOU SHOULD NOW KNOW

how to generate an output data set using PROC REG

Multiple Linear Regression

MODULES NEEDED   1, 2, 5, 9, 15, 16

The previous two modules dealt with simple linear regression. Often in data analysis, one independent variable does not account for enough of the variability in Y to be an adequate model. Other variables may also be helpful in predicting Y, including polynomial terms. This leads to multiple regression analysis.

In this module, you will learn how to use PROC REG to analyze multiple regression models, including polynomial models.

MULTIPLE REGRESSION

The model for multiple regression has the form

$$Y = \beta_o + \beta_1 x_1 + \beta_2 x_2 + \cdots + \beta_n x_n + \varepsilon$$

where x_i may be a completely different variable, a polynomial function of one of the variables (for example, x^2), or an interaction between two or more variables (for example, $x_1 * x_2$).

The theoretical multiple regression model is reflected in PROC REG's MODEL statement. In order to include polynomial functions or interactions, you first need to create new variables in the DATA step that represent these quantities. Then you include those new variables in the MODEL statement.

EXAMPLE 17.1

In the previous two modules, we found that a simple linear regression model was not appropriate for the DNase data. From the scatterplot, it seems as though a quadratic model might fit. Some of the output is shown in Figure 17.1.

SAS Program

```
filename rats 'assay.dat';

data protein;
   infile rats;
```

```
   input concen 1-10 absorb 12-16;
   concen = log(concen);
   concen2 = concen**2;

   label concen  = 'Ln of concentration of DNase'
         concen2 = 'Square of Ln of concentration of DNase'
         absorb  = 'Optical Density'
            ;
run;

proc reg data=protein;
   model absorb = concen concen2;
   plot absorb*p.='+';
title 'Quadratic Model for DNase Data';
run;
```

New Statements

```
model absorb = concen concen2;
```
This model contains the squared term CONCEN2.

Discussion of Output

The fitted quadratic model (see Figure 17.1) is ABSORB = 0.520 + 0.354*CONCEN + 0.067*CONCEN2. The F-test on the first page tests whether $\beta_1=\beta_2=0$. The p-value is 0.0001, indicating that the model is useful in explaining the variability in ABSORB.

Parameter estimates and standard errors of the estimates, along with t-tests and p-values are also given. You can see that the null hypothesis that $\beta_i=0$ is rejected for each of the βs in the model.

Although PROC UNIVARIATE showed that the residuals from the quadratic model are normally distributed (p-value=0.2532), the plot of the standardized residuals vs. y-hat on page 4 of Figure 17.1 still shows that there are problems with the model assumptions. Notice that the scatter is wider for larger y-hat values.

Figure 17.1 Partial Output Generated from Quadratic Model for Example 17.1

```
                      Quadratic Model for DNase Data                    1
                                              8:34 Tuesday, June 29, 1993

Model: MODEL1
Dependent Variable: ABSORB      Optical Density

                        Analysis of Variance

                          Sum of         Mean
       Source       DF    Squares       Square      F Value      Prob>F

       Model         2    14.31584      7.15792     1784.898     0.0001
       Error        39     0.15640      0.00401
       C Total      41    14.47224

              Root MSE       0.06333     R-square       0.9892
              Dep Mean       0.81874     Adj R-sq       0.9886
              C.V.           7.73467

                        Parameter Estimates

                    Parameter      Standard    T for H0:
       Variable  DF   Estimate        Error    Parameter=0    Prob > |T|

       INTERCEP   1   0.519573    0.01441067      36.055         0.0001
       CONCEN     1   0.354102    0.00878325      40.316         0.0001
       CONCEN2    1   0.066542    0.00587112      11.334         0.0001

                    Variable
       Variable  DF  Label

       INTERCEP   1  Intercept
       CONCEN     1  Ln of concentration of DNase
       CONCEN2    1  Square of Ln of concentration of DNase
```

Figure 17.1 cont.   Partial Output Generated from Quadratic Model for Ex. 17.1

```
                      Quadratic Model for DNase Data                    4
                                             8:34 Tuesday, June 29, 1993

              Plot of RESID*YHAT.   Legend: A = 1 obs, B = 2 obs, etc.

         |
       3 +
         |
         |
         |                                                A
         |
       2 +                                                A
   S     |
   t     |                                         A
   u     |                                                           A
   d     |                                                           A
   e     |                                  A             A
   n   1 +
   t     |
   i     |    B
   z     |    B
   e     |    A
   d     |                                  A             A
       0 +    A      A            A
   R     |              A       B       B       B
   e     |       C      A       A
   s     |              A       A       A
   i     |    A         A
   d     |              B       A
   u  -1 +
   a     |
   l     |
         |
         |                                                           A
      -2 +                                                           B
         |                                                           A
         |
         |
         |
         |
      -3 +
         |
         ---+-------------+-------------+-------------+-------------+--
          0.0           0.5           1.0           1.5           2.0
                        Predicted Value of ABSORB
```

PROBLEMS

The files for these problems are described in detail in the Appendix.

17.1   Use the file gas.dat for the following multiple regression models. Comment on model usefulness and whether model assumptions have been met.

      a)   Let mileage be the dependent variable and horsepower, car weight, and torque be the independent variables.

      b)   Let mileage be the dependent variable and horsepower, displacement, and torque be the independent variables.

17.2   Use the file grades.dat, where final exam grade is the dependent variable and quiz, first exam, second exam, and computer grades are the independent variables. Check the model assumptions.

17.3   Use the file electric.dat, where peak load is the dependent variable and house size, income, air capacity, and number in family are the independent variables. Check the model assumptions. Test whether family number should be in the model.

YOU SHOULD NOW KNOW

how to analyze multiple regression models using PROC REG

Multiple Regression:
Choosing the Best Model
MODULES NEEDED   1, 2, 5, 9, 15, 16, 17

If you want to choose the best regression model when there are several independent variables, it can take a lot of time to run all the possible regressions. Fortunately, PROG REG provides several options to help in this matter.

Stepwise regression is a method that performs regressions in stages. Independent variables are added and deleted from a model based on how significant each one is in the model. Forward regression only adds variables to the model, never deleting any. Backward regression starts with all the independent variables in the model and one at a time removes those that are not significant.

To help decide which multiple regression model is best, one can use several statistics: R^2, adjusted R^2, Mallow's C_P, and MSE.

In this module, you will learn how to use PROC REG to do forward, backward, and stepwise regression and how to generate several statistics to help in deciding which model is best.

PROC REG SELECTION= OPTION

The SELECTION= option in the MODEL statement in PROC REG is used to choose forward, backward, or stepwise regression. (See Module 15 for a description of this option.) Deciding whether to include a variable in the model or not is based on how significant it is measured by the p-value from the hypothesis test $H_o: \beta_i=0$. Whether a term enters or leaves the model can also be influenced by any correlation among the independent variables.

It is possible to set the significance level for a variable to enter the model with SLE= and to stay in the model with SLS=. The default for SLE is 0.5 for forward regression and 0.15 for stepwise regression. The default for SLS is 0.10 for backward selection and 0.15 for stepwise regression.

The option SELECTION=RSQUARE with CP, ADJRSQ, and MSE computes statistics for all possible models. CP requests Mallow's C_P statistic, which should be equal to the number of βs in the model; ADJRSQ requests the adjusted R^2; and MSE requests the mean squared error. Using this information, one can select the "best" multiple regression model.

EXAMPLE 18.1

It would be nice to find a "good" model that will predict peak hour electric load. Possible independent variables are house size, family income, air conditioning capacity, appliance index, and the number of family members. It is decided to use stepwise regression to help choose the model. The output is shown in Figure 18.1.

SAS Program

```
filename electric 'electric.dat';

data peak;
   infile electric;
   input housize 1-3 income 6-11 aircapac 14-16 applindx 19-23
         family 26-28 peak 31-35;

   label housize  = 'House Size'
         income   = 'Family Income'
         aircapac = 'Air Conditioning Capacity'
         applindx = 'Appliance Index'
         family   = 'Number of Family Members'
         peak     = 'Peak Hour Electric Load'
         ;
run;

proc reg data=peak;
   model peak = housize income aircapac applindx family
               / selection=stepwise;
title 'Multiple Regression Models Using Stepwise Regression';
run;
```

New Statements

```
model peak = housize income aircapac applindx family
/ selection=stepwise;
```
> In this stepwise regression analysis, the independent terms will enter and leave the model based on the default values of SLE and SLS.

Discussion of Output

Look at the output in Figure 18.1. Notice that SELECTION=STEPWISE produced six steps and a summary of the steps. In step 1, INCOME entered the model; in step 2, AIRCAPAC entered; in step 3, APPLINDX entered; in step 4, INCOME left the model; in step 5, HOUSIZE entered; and in step 6, FAMILY entered resulting in the model

PEAK = 0.00946 + 0.4147*HOUSIZE + 0.4433*AIRCAPAC + .3741*APPLINDX
+ 0.0456*FAMILY.

Figure 18.1 Output from Stepwise Regression of Example 18.1

```
           Multiple Regression Models Using Stepwise Regression         1
                                          10:06 Wednesday, June 30, 1993


                  Stepwise Procedure for Dependent Variable PEAK

Step 1    Variable INCOME Entered    R-square=0.86496506  C(p)=176.12081416

                 DF       Sum of Squares      Mean Square       F    Prob>F

Regression        1        104.30540566     104.30540566    371.52  0.0001
Error            58         16.28374893       0.28075429
Total            59        120.58915458

                 Parameter      Standard       Type II
Variable         Estimate         Error    Sum of Squares      F    Prob>F

INTERCEP       -1.58722364     0.33013225      6.48973189    23.12  0.0001
INCOME          0.25422641     0.01318956    104.30540566   371.52  0.0001

Bounds on condition number:              1,             1
-------------------------------------------------------------------------

Step 2    Variable AIRCAPAC Entered  R-square=0.94406093  C(p)=42.15750831

                 DF       Sum of Squares      Mean Square       F    Prob>F

Regression        2        113.84350910      56.92175455    480.98  0.0001
Error            57          6.74564549       0.11834466
Total            59        120.58915458

                 Parameter      Standard       Type II
Variable         Estimate         Error    Sum of Squares      F    Prob>F

INTERCEP       -0.09783964     0.27104248      0.01542068     0.13  0.7195
INCOME          0.14112188     0.01523337     10.15652373    85.82  0.0001
AIRCAPAC        0.40005041     0.04456132      9.53810344    80.60  0.0001

Bounds on condition number:     3.164529,      12.65812
-------------------------------------------------------------------------

Step 3    Variable APPLINDX Entered  R-square=0.96044539  C(p)=15.99313687

                 DF       Sum of Squares      Mean Square       F    Prob>F

Regression        3        115.81929724      38.60643241    453.25  0.0001
Error            56          4.76985734       0.08517602
Total            59        120.58915458

                 Parameter      Standard       Type II
Variable         Estimate         Error    Sum of Squares      F    Prob>F
```

Figure 18.1 cont.   Output from Stepwise Regression of Example 18.1

```
          Multiple Regression Models Using Stepwise Regression            2
                                            10:06 Wednesday, June 30, 1993


INTERCEP      0.16336521     0.23625295     0.04072697       0.48    0.4921
INCOME        0.03172332     0.02613347     0.12551066       1.47    0.2299
AIRCAPAC      0.45058091     0.03923324    11.23452424     131.90    0.0001
APPLINDX      0.39798578     0.08263348     1.97578815      23.20    0.0001

Bounds on condition number:        12.94024,        74.4442
----------------------------------------------------------------------------


Step 4    Variable INCOME Removed   R-square=0.95940457   C(p)=15.78226027

                 DF       Sum of Squares      Mean Square        F     Prob>F

Regression        2         115.69378658     57.84689329     673.55    0.0001
Error            57           4.89536800      0.08588365
Total            59         120.58915458

                 Parameter        Standard         Type II
Variable          Estimate          Error    Sum of Squares       F     Prob>F

INTERCEP        0.35049136     0.17977346      0.32644797       3.80    0.0561
AIRCAPAC        0.48041909     0.03070527     21.02446999     244.80    0.0001
APPLINDX        0.48517056     0.04103326     12.00680122     139.80    0.0001

Bounds on condition number:        2.070414,        8.281655
----------------------------------------------------------------------------


Step 5    Variable HOUSIZE Entered   R-square=0.96595833   C(p)= 6.51656442

                 DF       Sum of Squares      Mean Square        F     Prob>F

Regression        3         116.48409814     38.82803271     529.68    0.0001
Error            56           4.10505644      0.07330458
Total            59         120.58915458

                 Parameter        Standard         Type II
Variable          Estimate          Error    Sum of Squares       F     Prob>F

INTERCEP        0.20190296     0.17214178      0.10084257       1.38    0.2458
HOUSIZE         0.37051284     0.11284171      0.79031156      10.78    0.0018
AIRCAPAC        0.44795902     0.03004088     16.29980892     222.36    0.0001
APPLINDX        0.37915011     0.04979664      4.24965270      57.97    0.0001

Bounds on condition number:        3.637232,        28.59458
----------------------------------------------------------------------------
```

Figure 18.1 cont.   Output from Stepwise Regression of Example 18.1

```
                Multiple Regression Models Using Stepwise Regression        3
                                        10:06 Wednesday, June 30, 1993

Step 6    Variable FAMILY Entered    R-square=0.96858515    C(p)= 4.00114703

                DF      Sum of Squares    Mean Square        F      Prob>F

Regression       4       116.80086389    29.20021597     423.94    0.0001
Error           55         3.78829069     0.06887801
Total           59       120.58915458

               Parameter      Standard       Type II
Variable       Estimate       Error  Sum of Squares        F      Prob>F

INTERCEP       0.00945991     0.18946289     0.00017171     0.00     0.9604
HOUSIZE        0.41471461     0.11130668     0.95617367    13.88     0.0005
AIRCAPAC       0.44326697     0.02920181    15.87054524   230.42     0.0001
APPLINDX       0.37411600     0.04832677     4.12779114    59.93     0.0001
FAMILY         0.04561393     0.02127006     0.31676575     4.60     0.0364

Bounds on condition number:     3.766385,        42.9339
-----------------------------------------------------------------------

All variables in the model are significant at the 0.1500 level.
No other variable met the 0.1500 significance level for entry into the
model.

       Summary of Stepwise Procedure for Dependent Variable PEAK

        Variable        Number  Partial   Model
Step    Entered Removed   In     R**2      R**2      C(p)          F      Prob>F

  1     INCOME            1     0.8650    0.8650   176.1208    371.5185    0.0001
  2     AIRCAPAC          2     0.0791    0.9441    42.1575     80.5960    0.0001
  3     APPLINDX          3     0.0164    0.9604    15.9931     23.1965    0.0001
  4             INCOME    2     0.0010    0.9594    15.7823      1.4735    0.2299
  5     HOUSIZE           3     0.0066    0.9660     6.5166     10.7812    0.0018
  6     FAMILY            4     0.0026    0.9686     4.0011      4.5989    0.0364
```

EXAMPLE 18.2

For the same data as in Example 18.1, it is desired to see Mallow's C_p statistic, the adjusted R^2, R^2, and MSE to help determine which model is "best." The output is shown in Figure 18.2.

SAS Program

```
filename electric 'electric.dat';
```

```
data peak;
   infile electric;
   input @1 housize 3. @6 income 5. @14 aircapac 3.
         @19 applindx 5. @26 family 3. @31 peak 5.;

   label housize  = 'House Size'
         income   = 'Family Income'
         aircapac = 'Air Conditioning Capacity'
         applindx = 'Appliance Index'
         family   = 'Number of Family Members'
         peak     = 'Peak Hour Electric Load'
         ;
run;

proc reg data=peak;
   model peak = housize income aircapac applindx family
                / selection=rsquare cp adjrsq mse;
title 'Multiple Regression Models Using Stepwise Regression';
run;
```

New Statements

```
model peak = housize income aircapac applindx family
                / selection=rsquare cp adjrsq mse;
```
This analysis calls for the RSQUARE option to produce statistics helpful in deciding which independent variables should be included in the model.

Discussion of Output

In looking at Figure 18.2, you can see the statistics are given for all possible models. One choice for "best" is the model containing HOUSIZE, AIRCAPAC, APPLINDX, and FAMILY. For this model the R^2 and adjusted R^2 values are high, $C_P=4$, which is one less than the number of βs in the model, and MSE is small.

Figure 18.2 Output from RSQUARE Option in Example 18.2

```
         Multiple Regression Models Using the RSQUARE Option          1
                                    10:06 Wednesday, June 30, 1993


N = 60       Regression Models for Dependent Variable: PEAK


        R-square      Adj      C(p)       MSE   Variables in Model
    In                Rsq


     1  0.864965   0.862637    176.1   0.28075  INCOME
     1  0.859837   0.857420    184.9   0.29142  AIRCAPAC
     1  0.785057   0.781351    313.5   0.44689  APPLINDX
     1  0.732112   0.727493    404.5   0.55697  HOUSIZE
     1  0.004482  -.012682    1655.3   2.06981  FAMILY
   ------------------------------------------------------------------
     2  0.959405   0.957980   15.7823  0.08588  AIRCAPAC APPLINDX
     2  0.944061   0.942098   42.1575  0.11834  INCOME AIRCAPAC
     2  0.930718   0.928287   65.0943  0.14657  HOUSIZE AIRCAPAC
     2  0.876680   0.872353    158.0   0.26090  HOUSIZE INCOME
     2  0.867282   0.862625    174.1   0.28078  INCOME APPLINDX
     2  0.865118   0.860386    177.9   0.28536  INCOME FAMILY
     2  0.859866   0.854949    186.9   0.29647  AIRCAPAC FAMILY
     2  0.830790   0.824853    236.9   0.35798  HOUSIZE APPLINDX
     2  0.786744   0.779261    312.6   0.45116  APPLINDX FAMILY
     2  0.742486   0.733451    388.7   0.54480  HOUSIZE FAMILY
   ------------------------------------------------------------------
     3  0.965958   0.964135    6.5166  0.07330  HOUSIZE AIRCAPAC APPLINDX
     3  0.960656   0.958548   15.6312  0.08472  AIRCAPAC APPLINDX FAMILY
     3  0.960445   0.958326   15.9931  0.08518  INCOME AIRCAPAC APPLINDX
     3  0.953174   0.950665   28.4933  0.10084  HOUSIZE INCOME AIRCAPAC
     3  0.944322   0.941339   43.7096  0.11990  INCOME AIRCAPAC FAMILY
     3  0.934355   0.930838   60.8418  0.14136  HOUSIZE AIRCAPAC FAMILY
     3  0.878239   0.871716    157.3   0.26220  HOUSIZE INCOME FAMILY
     3  0.877710   0.871159    158.2   0.26334  HOUSIZE INCOME APPLINDX
     3  0.867612   0.860520    175.6   0.28508  INCOME APPLINDX FAMILY
     3  0.836977   0.828243    228.2   0.35105  HOUSIZE APPLINDX FAMILY
   ------------------------------------------------------------------
     4  0.968585   0.966300    4.0011  0.06888  HOUSIZE AIRCAPAC APPLINDX
                                                FAMILY
     4  0.966036   0.963565    8.3837  0.07447  HOUSIZE INCOME AIRCAPAC
                                                APPLINDX
     4  0.961505   0.958705   16.1715  0.08440  INCOME AIRCAPAC APPLINDX
                                                FAMILY
     4  0.954787   0.951499   27.7193  0.09913  HOUSIZE INCOME AIRCAPAC
                                                FAMILY
     4  0.879506   0.870742    157.1   0.26419  HOUSIZE INCOME APPLINDX
                                                FAMILY
   ------------------------------------------------------------------
     5  0.968586   0.965677    6.0000  0.07015  HOUSIZE INCOME AIRCAPAC
                                                APPLINDX FAMILY
   ------------------------------------------------------------------
```

PROBLEMS

The files for these problems are described in detail in the Appendix.

18.1   Use the file gas.dat with gas mileage as the dependent variable and displacement, horsepower, torque, transmission speeds, car weight, and car length as the independent variables to find the best regression model using

    a)   stepwise regression.
    b)   backward regression.
    c)   forward regression.
    d)   the RSQUARE option and C_p, adjusted R^2, R^2, and MSE.

18.2   Use the file gas.dat with horsepower as the dependent variable and compression ratio, rear axle ratio, # carburetor barrels, car weight, and torque as the independent variables to find the best regression model using

    a)   stepwise regression.
    b)   backward regression.
    c)   forward regression.
    d)   the RSQUARE option and C_p, adj R^2, R^2, and MSE.

18.3   Use the file grades.dat with final exam grade as the dependent variable and the other grades as independent variables to find the best regression model using

    a)   stepwise regression.
    b)   backward regression.
    c)   forward regression.
    d)   the RSQUARE option and C_p, adjusted R^2, R^2, and MSE.

YOU SHOULD NOW KNOW

how to do stepwise, forward, and backward regression

how to generate statistics used in comparing regression models

---

Chi-Square Tests

MODULES NEEDED 1, 2, 5, 7

When variables are categorical, hypotheses generally involve counts and proportions. Two-way tables are a good way to view results when there are two variables. In this module you will learn how to generate statistics for analyzing two-way tables using PROC FREQ. PROC FREQ's general form is described in Module 7.

CHI-SQUARE TESTS

The CHISQ option on the TABLES statement in PROC FREQ requests several statistics relating to two-way tables. The usual test described in introductory statistics texts is the one labeled Chi-Square on the output. This statistic is used for tests of independence and tests of homogeneity.

The EXPECTED option computes and prints the expected counts for each cell in the table.

EXAMPLE 19.1

Debate coach Ronda Nielson wanted to know if Granger, Hunter, Kearns, and Skyline high schools differed in the effectiveness of debate classes in teaching valuable life skills. A χ^2 test was in order. This is a test of homogeneity. The output is shown in Figure 19.1.

SAS Program

```
filename debate 'debate.dat';

data one;
   infile debate;
   input id school gender compare argue research reason speak;
   if school=3 or school=5 or school=6 or school=8;

   label id        = 'Survey Number'
         school    = 'High School'
         gender    = '1=female    2=male'
         compare   = 'How Debate Compares to Other Classes'
         argue     = 'Argumentation'
```

```
            research = 'Research'
            reason   = 'Reasoning'
            speak    = 'Speaking'
            ;
run;

proc freq data=one;
   tables school*compare /chisq expected;
title 'Comparing Schools in the Debate Survey';
run;
```

Discussion of Output

Notice in Figure 19.1 that the expected counts are included as the second number in each cell. The χ^2 test statistic has a value of 0.343 and a p-value of 0.342, indicating that the high schools are not significantly different from one another regarding the response to the question about comparing debate to other classes.

Notice the SAS WARNING message. This refers to the rule of thumb that expected cell counts should be five or more in order to do a χ^2 test. Some books may give a different rule of thumb.

Figure 19.1 Output for the Chi-Square Test in Example 19.1

```
                 Comparing Schools in the Debate Survey           1
                          13:33 Wednesday, June 30, 1993


                      TABLE OF SCHOOL BY COMPARE

     SCHOOL(High School)
                 COMPARE(How Debate Compares to Other Classes)
     Frequency|
     Expected |
     Percent  |
     Row Pct  |
     Col Pct  |        1|        2|  Total
     ---------+--------+--------+
            3 |     17 |      6 |     23
              | 19.401 | 3.5986 |
              |  11.56 |   4.08 |  15.65
              |  73.91 |  26.09 |
              |  13.71 |  26.09 |
     ---------+--------+--------+
            5 |     22 |      3 |     25
              | 21.088 | 3.9116 |
              |  14.97 |   2.04 |  17.01
              |  88.00 |  12.00 |
              |  17.74 |  13.04 |
     ---------+--------+--------+
            6 |     30 |      7 |     37
              | 31.211 | 5.7891 |
              |  20.41 |   4.76 |  25.17
              |  81.08 |  18.92 |
              |  24.19 |  30.43 |
     ---------+--------+--------+
            8 |     55 |      7 |     62
              | 52.299 | 9.7007 |
              |  37.41 |   4.76 |  42.18
              |  88.71 |  11.29 |
              |  44.35 |  30.43 |
     ---------+--------+--------+
     Total         124       23      147
                 84.35    15.65   100.00

     Frequency Missing = 1
```

Figure 19.1 cont.   Output for the Chi-Square Test in Example 19.1

```
Comparing Schools in the Debate Survey              2
                              13:33 Wednesday, June 30, 1993


              STATISTICS FOR TABLE OF SCHOOL BY COMPARE

       Statistic                    DF    Value      Prob
       -------------------------------------------------------
       Chi-Square                    3    3.343      0.342
       Likelihood Ratio Chi-Square   3    3.167      0.367
       Mantel-Haenszel Chi-Square    1    2.172      0.141
       Phi Coefficient                    0.151
       Contingency Coefficient            0.149
       Cramer's V                         0.151

       Effective Sample Size = 147
       Frequency Missing = 1
       WARNING:  25% of the cells have expected counts less
                 than 5. Chi-Square may not be a valid test.
```

EXAMPLE 19.2

Ms. Nielson lost the original data file, but she did have a summary table of counts as follows:

	More Effective	Just as Effective
Granger	17	6
Hunter	22	3
Kearns	30	7
Skyline	55	7

SAS Program

```
data one;
   input school $ compare count;

   label school   = 'High School'
         compare  = 'How Debate Compares to Other Classes'
         ;
cards;
granger 1 17
granger 2  6
hunter  1 22
hunter  2  3
kearns  1 30
kearns  2  7
skyline 1 55
skyline 2  7
```

```
;
run;

proc freq data=one;
   tables school*compare /chisq expected;
   weight count;
title 'Comparing Schools in the Debate Survey';
run;
```

New Statements

```
weight count;
```

> This statement tells SAS that the data should be analyzed as if there were multiple observations with frequency COUNT.

PROBLEMS

The files for these problems are described in detail in the Appendix.

19.1   Use the file debate.dat to see if Granger, Hunter, Kearns, and Skyline high schools differ in the effectiveness of speech/debate in teaching

a)   research skills. (Combine somewhat effective and not at all effective into one category.)
b)   reasoning skills.
c)   speaking skills.
d)   argumentation skills.

19.2   Use the file debate.dat to see if at Skyline High School gender is related to responses in the following areas:

a)   Compared to other classes.
b)   Argumentation.
c)   Research.
d)   Reasoning.
e)   Speaking.

19.3   Use the file src.dat to test if

a)   gender is related to environmentalist (make three categories 1–3 and 4–7 and 8–10), general quality of environment, plants and animals exist primarily to be used by humans, political party, and liberal-conservative.

b)   political party (Republicans, Democrats, and Independents only) is related to environmentalist (coded as in (a)), air quality, health hurt by lack of environmental quality, plants and animals exist primarily to be used by humans, we must protect environment even if jobs are lost, and we must protect environment even if population growth slows.

c)   liberal-conservative is related to environmentalist (coded as in (a)), air quality, health hurt by lack of environmental quality, plants and animals exist primarily to be used by humans, we must protect environment even if population growth slows, and we must protect environment even if jobs are lost.

YOU SHOULD NOW KNOW

how to use PROC FREQ to analyze two-way tables

Nonparametric One-Way ANOVA

MODULES NEEDED 1, 2, 5, 12

In analysis of variance, when it cannot be assumed that the samples came from a normal distribution, nonparametric tests are in order. The Wilcoxon rank-sum test is used for two groups, and the Kruskal-Wallis test is used for more than two groups.

In this module, you will learn how to do nonparametric one-way ANOVA using PROC NPAR1WAY.

PROC NPAR1WAY

PROC NPAR1WAY performs a nonparametric one-way analysis of variance. This is appropriate when the data is not normally distributed but is continuous.

The CLASS variable is used to distinguish the groups. It should have two possible values, which can be character or numeric. It works the same as the CLASS variable in PROC ANOVA and PROC GLM. See Module 12.

Ties in the data are handled by assigning average ranks, and variance estimates are adjusted.

PROC NPAR1WAY General Form

```
proc npar1way data=data set <options>;
    by variables;
    class variable;
    var variables;
```

PROC NPAR1WAY Option	Description
anova	This option performs a usual analysis of variance.
wilcoxon	This option performs a Wilcoxon rank-sum test if the variable has two levels and a Kruskal-Wallis test if the variable has more than two levels.

EXAMPLE 20.1

In Example 12.1, researchers wanted to test whether the average braking time of drivers of different types of trucks equipped with a center high-mounted stop lamp are the same or not. They are concerned that the data is not normally distributed and prefer to use a nonparametric test. The output is shown in Figure 20.1.

SAS Program

```
filename inbrakes 'taillite.dat';

data one;
   infile inbrakes;
   input id vehtype group positn speedzn resptime
     follotme folltmec;

   if group = 1;

   label vehtype = 'Vehicle Type'
         group   = 'Group - Light On=1    Light Off=2'
         positn  = 'Light Position'
         speedzn = 'Speed Zone'
         resptime= 'Response Time'
         follotme= 'Following Time in Video Frames'
         folltmec= 'Following Time in Categories'
         ;
run;

proc npar1way data=one wilcoxon;
   class vehtype;
   var resptime;
title 'Nonparametric One-Way ANOVA for Tail Light Study';
run;
```

New Statements

```
proc npar1way data=one wilcoxon;
class vehtype;
var resptime;
```
The variable VEHTYPE contains two values that divide the data into two groups. The WILCOXON option calls for a Kruskal-Wallis test because there are four different vehicle types.

Discussion of Output

The output (see Figure 20.1) shows the Kruskal-Wallis test statistic = 4.0213 with a p-value of 0.2592. The conclusion is that the vehicle types are not significantly different. The Sum of Scores is the sum of the ranks for each group.

Figure 20.1 Output from NPAR1WAY for Example 20.1

```
Nonparametric One-Way ANOVA for Tail Light Study        1
                          8:17 Thursday, July 1, 1993


              N P A R 1 W A Y   P R O C E D U R E


    Wilcoxon Scores (Rank Sums) for Variable RESPTIME
            Classified by Variable VEHTYPE



                  Sum of     Expected     Std Dev      Mean
 VEHTYPE     N    Scores     Under H0     Under H0     Score

 1         157  55802.0000   57619.0   2351.16490  355.426752
 2         218  76140.5000   80006.0   2619.71401  349.268349
 3         189  72170.5000   69363.0   2506.98869  381.854497
 4         169  64898.0000   62023.0   2413.82025  384.011834
              Average Scores were used for Ties


    Kruskal-Wallis Test (Chi-Square Approximation)
      CHISQ= 4.0213  DF=  3  Prob > CHISQ=     0.2592
```


PROBLEMS

The files for these problems are described in detail in the Appendix.

20.1   Use the file taillite.dat to answer the following questions:

a)   For the experimental group in the 30 mph speed zone, are the average response times for truck types different?

b)   For the experimental group and minivans, are the average response times for speed zones different?

20.2   Use the file wine.dat to answer the following questions:

a)   Do the average ratings differ for wine brand?

b)   Do the average ratings differ for temperature?

20.3   Use the file calls.dat to answer the following questions:

a)   Are the number of calls different for different shifts?

b)   Are the nubmer of calls different for different days?

YOU SHOULD NOW KNOW

how to use PROC NPAR1WAY

Analysis of Covariance

MODULES NEEDED _1, 2, 5, 12, 13, 15_

The terms in an ANOVA model are qualitative, and the terms in a regression model are usually quantitative. When a model contains both qualitative and quantitative terms, one performs an analysis of covariance. It is a mixture of regression and ANOVA.

In ANOVA, the response variable may be related to a numeric variable that cannot be controlled. This relationship can explain some of the inherent variability in Y, and by including it in the model, it may be easier to find differences among the main effects. This extra variable is known as a covariate, a regressor, or a concomitant variable.

The model looks like this for two qualitative factors with no interaction and one covariate:

$$Y_{ijk} = \mu + A_i + B_j + \beta X_{ijk} + \varepsilon_{ijk}$$

In this module, you will learn how to do analysis of covariance using PROC GLM.

USING PROC GLM FOR ANCOVA

The CLASS statement PROC GLM tells SAS which variables are main effects. Any variables in the MODEL statement not included in the CLASS statement are considered regressors or covariates.

The SOLUTION option in the MODEL statement requests a least-squares solution for β. PROC GLM's default in ANCOVA is not to compute β-hat.

The Type I Sums of Squares on the output provide F-tests for an ANOVA model that does not include any covariates. The Type III Sums of Squares are the correct ones to use for testing whether each factor is significant in the ANCOVA model.

EXAMPLE 21.1

In analyzing the data in the center high-mounted stop lamp (CHMSL) study, following time was added to the model as a covariate. The ANCOVA output is shown in Figure 21.1.

SAS Program

```
filename datain 'taillite.dat';

data lights;
   infile datain;
   input id vehicle group position speedzn resptime
         folltime folltimc;

   label id       = 'Vehicle ID'
         vehicle  = 'Vehicle Type'
         group    = 'Experimental (1) or Control (2)'
         position = 'CHMSL high(1) or low (2)'
         speedzn  = 'Speed Zone'
         resptime = 'Response Time'
         folltime = 'Following Time'
         folltimc = 'Following Time in Categories'
         ;
run;

proc glm data=lights;
   class vehicle group position speedzn;
   model resptime = group vehicle position
         speedzn folltime / solution;
title 'ANCOVA for Tail Light Study';
run;
```

New Statements

```
class vehicle group position speedzn;
model resptime = group vehicle position speedzn folltime
/solution;
```

> The variable FOLLTIME is a covariate because it is not listed in the CLASS statement. /SOLUTION calls for SAS to print the estimate of β.

Discussion of Output

Look at the output in Figure 21.1. The Type III Sums of Squares show that the factors VEHICLE and SPEEDZN are significant (p-value=0.0001). POSITION is not significant, and GROUP has a p-value of 0.0431. The F-statistic for testing $H_o{:}\beta{=}0$ is 121.98 (p-value=0.0001).

The estimate for β is 0.558. Note that the square of the t-statistic equals the F-statistic in the Type III Sums of Squares section.

Figure 21.1 ANCOVA Output for Example 21.1

```
               ANCOVA for Tail Light Study                  1
                       8:17 Thursday, July 1, 1993

        General Linear Models Procedure
          Class Level Information

        Class     Levels    Values

        VEHICLE      4     1 2 3 4

        GROUP        2     1 2

        POSITION     2     1 2

        SPEEDZN      3     30 40 50


   Number of observations in data set = 1087
```

Figure 21.1 cont.   ANCOVA Output for Example 21.1

```
                    ANCOVA for Tail Light Study            2
                   General Linear Models Procedure

Dependent Variable: RESPTIME   Response Time

Source                  DF   Sum of Squares   F Value   Pr > F

Model                    8   61727.6839237     24.36   0.0001
Error                 1078  341441.1422032
Corrected Total       1086  403168.8261270

              R-Square             C.V.      RESPTIME Mean
              0.153106          39.40806        45.16099356


Source                  DF      Type I SS   F Value   Pr > F

GROUP                    1    1458.2719421      4.60   0.0321
VEHICLE                  3    9189.0328273      9.67   0.0001
POSITION                 1     228.2632041      0.72   0.3961
SPEEDZN                  2   12217.6060241     19.29   0.0001
FOLLTIME                 1   38634.5099261    121.98   0.0001

Source                  DF    Type III SS   F Value   Pr > F

GROUP                    1    1298.5804370      4.10   0.0431
VEHICLE                  3    7842.1149769      8.25   0.0001
POSITION                 1     277.8748438      0.88   0.3491
SPEEDZN                  2   20601.4716561     32.52   0.0001
FOLLTIME                 1   38634.5099261    121.98   0.0001


                          T for H0:   Pr > |T|  Std Error of
Parameter        Estimate  Parameter=0           Estimate

INTERCEPT      35.21344235 B    14.93   0.0001    2.35793916
GROUP       1  -2.33383802 B    -2.02   0.0431    1.15261649
            2   0.00000000 B      .        .          .
VEHICLE     1  -7.54779173 B    -4.62   0.0001    1.63358324
            2  -5.48046524 B    -3.62   0.0003    1.51399016
            3  -3.14022945 B    -2.03   0.0424    1.54553980
            4   0.00000000 B      .        .          .
POSITION    1   1.01240957 B     0.94   0.3491    1.08088670
            2   0.00000000 B      .        .          .
SPEEDZN    30  -7.61749702 B    -5.36   0.0001    1.42189379
           40 -10.20322488 B    -7.85   0.0001    1.30019413
           50   0.00000000 B      .        .          .
FOLLTIME       0.55833761       11.04   0.0001    0.05055426

NOTE: The X'X matrix has been found to be singular and a generalized
inverse was used to solve the normal equations. Estimates followed by the
letter 'B' are biased, and are not unique estimators of the parameters.
```

PROBLEMS

The files for these problems are described in detail in the Appendix.

21.1   Use the file gas.dat to do ANCOVA, where gas mileage is the dependent variable and transmission type and transmission speeds are the factors and the covariate is

      a)   car weight.
      b)   torque.

21.2   Use the file dummy.dat to do ANCOVA, where stiffness1 is the dependent variable and impactor and species are the factors and the covariate is

      a)   % calcium.
      b)   % magnesium.

YOU SHOULD NOW KNOW

how to use SAS to do ANCOVA

Description of Data Sets

assay.dat

These data are taken from data generated during development of an ELISA assay for the recombinant protein DNase in rat serum [Marie Davidian and David Giltinan, "Some Simple Methods for Estimating Intraindividual Variability in Nonlinear Mixed Effects Models," *Biometrics* (Mar 1993), Vol 59, 59–73]. The independent variable is concentration of DNase, and the dependent variable is optical density, a measure of absorbance. The main use of the model was to calibrate unknown samples. The authors suggest using ln concentration as the independent variable.

Number of observations = 48
Used in Modules 15, 16, 17

variable	type	columns	additional description
concentration	numeric	1–10	in ng/ml
optical density	numeric	12–16	

```
----+----1----+----2----+----3----+----4----+----5----+----6
1.5625      .658
1.5625      .644
3.125       .994
```

bonescor.dat

Dual-photon absorptiometry (DPA) is a method to measure bone mineral density and evaluate the quality of bone being considered for transplant. This technique is expensive and often unavailable. An alternative method is to use a bone score based on the Singh index, CC ratio, calcar width, and cortical shaft index (CSI), all measures of bone quality. Data was collected by Renee Smith of the Bone and Joint Research Lab, VA Medical Center, Salt Lake City, Utah, to compare the two methods. Results from the study appear in "Roentgenographic Procedure for Selecting Proximal Femur Allograft for Use in Revision Arthroplasty," *The Journal of Arthroplasty* (1993), Vol 8 No 4, 347–360. Data used with permission.

Number of observations = 30
Used in Modules 14, 15, 16

variable	type	columns	additional description
Singh index	numeric	1	high values are better
CC ratio	numeric	3–5	low values are better
CSI	numeric	7–9	high values are better
calcar width	numeric	11–12	high values are better
bone score	numeric	14–15	
% average young normal	numeric	17–21	DPA values on three measures were compared to normal individuals aged 20–40 and averaged to get % young normal

```
----+----1----+----2----+----3----+----4----+----5----+----6
5 .47 .62  9 10 125.7
5 .57 .67 10 10 135.6
5 .50 .65  7 10 106.8
```

brownie.dat

These data were collected by Becki Graf, a student in a design of experiments class, as a class project. She wanted to know if the type of pan (glass or aluminum) or brand of brownie mix (Betty Crocker or Gold Medal) affected the amount of hard crust on brownies. The experiment was conducted on two different days.

Number of observations = 8
Used in Module 13

variable	type	columns	additional description
day	numeric	1	1=first day 2=second day
pan type	alphanumeric	3–7	glass or alum
mix brand	alphanumeric	9–13	betty or gold
width of hard crust	numeric	15–17	in mm

```
----+----1----+----2----+----3----+----4----+----5----+----6
1 glass betty 1.4
1 glass gold  1.8
1 alum  betty 1.0
```

calls.dat

These data were collected by Dawn Wick, a student in a design of experiments class, as a class project. She wanted to know if the number of customer calls to a help line were different for different days of the week and different shifts. The experiment was conducted over 11 weeks.

Number of observations = 165
Used in Modules 12, 13, 20

variable	type	columns	additional description
week	numeric	1-2	
shift	numeric	4	
day	alphanumeric	6-8	
number of calls	numeric	10-12	

```
----+----1----+----2----+----3----+----4----+----5----+----6
 1 1  Mon    9
 1 2  Mon   16
 1 3  Mon   14
```

cataract.dat

These data were provided by Dr. Monte Leidenix from the University of Utah Medical Center Department of Opthamology. He wanted to compare the amount of astigmatism (0=no astigmatism) for two different lenses measured one month after cataract surgery.

Number of observations = 24
Used in Module   11

variable	type	columns	additional description
lens type	alphanumeric	1-2	NF=nonfolding F=folding
astigmatism	numeric	4-7	measured in diopters

```
----+----1----+----2----+----3----+----4----+----5----+----6
NF 1.03
NF 0.17
NF 1.20
```

china#1.dat

The article "Unravelling the Mysteries of China's Foreign Trade Regime" by Arvind Panagariya [*The World Economy* (Jan 93), Vol 16 No 1, 51–67] discusses China's foreign trade policies from 1955 to 1989. This data set is from the *Almanac of China's Foreign Economic Relations and Trade 1990/91* (Table 1, p. 53 in the paper).

Number of observations = 35
Used in Modules 2, 3, 5, 6, 9

variable	type	columns	additional description
year	numeric	1–4	
total	numeric	6–10	exports + imports ($ billions)
total exports	numeric	12–16	in billions of US dollars
total imports	numeric	18–22	in billions of US dollars

```
----+----1----+----2----+----3----+----4----+----5----+----6
1955  3.15  1.41  1.73
1956  3.21  1.65  1.56
1957  3.10  1.60  1.51
```

debate.dat

These data were taken from a survey of speech and debate students from high schools in the Granite School District, Salt Lake City, Utah. Data were collected by Ronda Nielson, debate teacher at Skyline High School, and analyzed by E&M Statistical Consulting. Questions asked about the effectiveness of speech and debate classes in teaching certain skills. Data used with permission.

Number of observations = 327
Used in Modules 10, 19

variable	type	columns	additional description
survey number	numeric	1–3	ID variable
school	numeric	5	1=Cottonwood 2=Cyprus 3=Granger 4=Granite 5=Hunter 6=Kearns 7=Olympus 8=Skyline 9=Taylorsville
gender	numeric	8	1=female 2=male
compared to other classes	numeric	11	how debate compares in teaching valuable skills 1=more effective 2=just as effective 3=less effective
argumentation	numeric	15	effectiveness of speech/debate in teaching 1=very effective 2=somewhat effective 3=not at all effective
research	numeric	18	same as for argumentation
reasoning	numeric	21	same as for argumentation
speaking	numeric	25	same as for argumentation

```
----+----1----+----2----+----3----+----4----+----5----+----6
1    6  1  1   1  1  1    1
108  7  1  1   1  1  1    2
56   3  1  1   1  1  1    1
```

dummy.dat

These data were taken from data provided by Scott McClelland, a graduate student at the University of Utah, from a study comparing the Hybrid III dummy to humans. Dummies and human skulls were hit with side impacts, and stiffness measurements were taken in two locations.

Number of observations = 24
Used in Modules 13, 21

variable	type	columns	additional description
species	alphanumeric	1	h=human d=dummy
impactor	alphanumeric	3–5	bar d1=disc
stiffness1	numeric	7–10	stiffness = force vs. displacement (in N/mm) at head location 1
stiffness2	numeric	12–15	stiffness at head location 2
calcium	numeric	17–20	% weight of calcium in bone
magnesium	numeric	22–26	% weight of magnesium in bone

```
----+----1----+----2----+----3----+----4----+----5----+----6
h bar 2586 2586 21.8 0.256
h bar 5769 5556 22.2 0.220
h bar 5000 4000 18.9 0.214
```

electric.dat

These data are for residential electric utility customers [Douglas Montgomery and David Friedman, "Prediction Using Regression Models with Multicollinear Predictor Variables," *IIE Transactions* (May 1993) Vol 25 No 3, 73–85; Montgomery, D. C. and Askin, R. G., "Problems of Nonnormality and Multicollinearity for Forecasting Methods Based on Least Squares," *AIIE Transactions* (1981), Vol 13, 102–115].

Number of observations = 60
Used in Modules 14, 15, 16, 17, 18

variable	type	columns	additional description
house size	numeric	1–3	in square feet / 1000
family income	numeric	6–11	in thousands of dollars
air conditioning capacity	numeric	14–16	in tons
appliance index	numeric	19–23	the connected load of all electric appliances
# family members	numeric	26–28	
peak hour load	numeric	31–35	

```
----+----1----+----2----+----3----+----4----+----5----+----6
3.2  34.990  7.0  7.789  4.0  7.518
1.3  14.160  0.5  3.652  4.0  2.349
4.1  22.962  3.0  5.854  1.0  5.059
```

gas.dat

These data are for foreign and domestic cars [Douglas Montgomery and David Friedman, "Prediction Using Regression Models with Multicollinear Predictor Variables," *IIE Transactions* (May 1993) Vol 25 No 3, 73–85; Montgomery, D. C. and Peck, E. A., *Introduction to Linear Regression Analysis, 2nd Edition*, John Wiley and Sons, New York (1192)].

Number of observations = 30
Used in Modules 11, 14, 17, 18, 21

variable	type	columns	additional description
displacement	numeric	1–5	in cubic inches
horsepower	numeric	7–9	in ft-lbs
torque	numeric	11–13	in ft-lbs
compression ratio	numeric	15–17	
rear axle ratio	numeric	19–21	
# carburetor barrels	numeric	23	
transmission speeds	numeric	25	how many
car length	numeric	27–31	in inches
car width	numeric	33–36	in inches
car weight	numeric	38–41	in lbs
transmission type	numeric	43	0=manual 1=automatic
gas mileage	numeric	45–48	in miles per gallon

```
----+----1----+----2----+----3----+----4----+----5----+----6
318.0 140 255 8.5 2.7 2 3 215.3 76.3 4370 1 19.7
440.0 215 330 8.2 2.9 4 3 184.5 69.0 4215 1 11.2
351.0 143 255 8.0 3.0 2 3 199.9 74.0 3890 1 18.3
```

grades.dat

These data contain grades for students in an introductory, calculus-based statistics class.

Number of observations = 49
Used in Modules 7, 11, 14, 17, 18

variable	type	columns	additional description
student ID	alphanumeric	1–3	
gender	alphanumeric	5	f=female m=male
class	numeric	7	2=sophomore 3=junior 4=senior
quiz grade	numeric	9–10	maximum points=50
first exam grade	numeric	12–14	maximum points=100
second exam grade	numeric	16–18	maximum points=100
computer lab grade	numeric	20–22	maximum points=100
final exam grade	numeric	25–27	maximum points=200

```
----+----1----+----2----+----3----+----4----+----5----+----6
air f 4 50   93   93   98   162
aln m 4 49   95   98   97   175
bam m 4 39   63   84   95    95
```

handinj.dat

These data were collected during a study of hand injuries at Beaumont Hospital in Dublin, Ireland. Dr. Michael E. O'Sullivan provided the data, which consists of soft tissue injuries of the hand. He was interested in determining if there was a difference in the number of days of work lost and cost incurred for injuries occurring during work versus injuries sustained during sports activity. The results of this study will appear in "The Economic Impact of Hand Injuries," *British Journal of Hand Surgery.*

Number of observations = 38
Used in Modules 2, 9, 11, 14

variable	type	columns	additional description
ID	alphanumeric	1–5	
injury type	alphanumeric	7–11	work or sport
days of work lost	numeric	13–14	
cost	numeric	16–19	in Irish pounds

```
----+----1----+----2----+----3----+----4----+----5----+----6
ID001 work    0    85
ID007 work    0   165
ID009 work    5   205
```

robot.dat

In a manufacturing scheduling study, the performance of robots trained with a hybrid method is compared to human performance (Yuehwen Yih et al., "Robot Scheduling in a Circuit Board Production Line: A Hybrid OR/ANN Approach," *IIE Transactions* (Mar 1993), Vol 25 No 2, 26–33). The response variables were product quality, the ratio of good jobs to total jobs, and throughput, the number of good jobs times product quality. Human and hybrid performance was evaluated for eight simulated manufacturing conditions. The data are paired.

Number of observations = 8
Used in Module   11

variable	type	columns	additional description
human throughput	numeric	1–5	
hybrid throughput	numeric	7–11	
human quality	numeric	13–16	
hybrid quality	numeric	18–21	

```
----+----1----+----2----+----3----+----4----+----5----+----6
185.4 180.4 .889 .997
146.3 248.5 .791 .959
174.4 185.5 .866 .821
```

sclero.dat

A randomized, double-blind, controlled study was conducted in 1989 to determine the safety and efficacy of a particular compound in the treatment of scleroderma, a multi-system skin disease characterized by thickening of the skin and possible involvement of the blood vessels and internal organs. Skin mobility is the sum of mobility scores from 20 skin locations and assesses the ability of the skin to be stretched, compressed, and lifted. Skin thickening is the sum of thickening scores graded 0–3 (3 is the worst) from 15 areas. Patient assessment is the sum of scores from the hand, forearms, and arms using a score of 0–3 (3 is the worst). Data used with permission.

Number of observations = 76
Used in Modules 2, 5, 6, 7, 10

variable	type	columns	additional description
clinic number	numeric	1–2	
patient ID	numeric	4–5	
placebo/drug	numeric	8	1=drug 2=placebo
skin thickening at 1st visit	numeric	11–12	the higher the number the worse the thickening
skin thickening at 2nd visit	numeric	15–16	the higher the number the worse the thickening
skin mobility at 1st visit	numeric	19–21	the higher the number the better the mobility
skin mobility at 2nd visit	numeric	24–26	the higher the number the better the mobility
patient assessment at 1st visit	numeric	29	the higher the number the worse the patient
patient assessment at 2nd visit	numeric	32	the higher the number the worse the patient

```
----+----1----+----2----+----3----+----4----+----5----+----6
1 04  1    .    .  526  535  2  3
2 05  1   30   28  227  260  6  2
2 06  2   18   21  375  409  6  5
```

src.dat

This is a subset of data provided by the University of Utah Survey Research Center from the 1992 Utah Consumer Survey. This telephone survey is conducted during the first month of each calendar quarter and asks many questions regarding economic conditions in the state of Utah. Complete survey results are contained in the report *Utah Consumer Survey January 1993 Report* by the University of Utah Survey Research Center.

Number of observations = 506
Used in Modules 10, 11, 19

variable	type	columns	additional description
ID	numeric	1–4	
gender	numeric	6	1=female 5=male
environmentalist	numeric	8–9	a measure from 1–10 of how strongly one identifies as an environmentalist (10=strongly identify)
general quality of environment	numeric	12	1=poor 3=fair 5=good 7=excellent
air quality	numeric	14	1=poor 3=fair 5=good 7=excellent
health hurt by lack of environmental quality?	numeric	16	1=definitely not 3=possibly 5=probably 7=definitely
plants and animals exist primarily to be used by humans	numeric	18–19	1=strongly disagree with statement 3=somewhat disagree 5=neutral 7=somewhat agree 9=strongly agree
we must protect the environment even if jobs are lost because of it	numeric	22–23	1=strongly disagree with statement 3=somewhat disagree 5=neutral 7=somewhat agree 9=strongly agree
we must protect the environment even if population growth slows	numeric	25–26	1=strongly disagree with statement 3=somewhat disagree 5=neutral 7=somewhat agree 9=strongly agree
employment status	numeric	30–31	1=working 3=keeping house 5=student 9=have job but not working temporarily 11=unemployed, laid off, looking for work 13=retired 15=disabled/unable to work
hours worked	numeric	35–36	during previous week
yearly income	numeric	40–41	1=<5,000 5=5,000–10,000 10=10,000–15,000 15=15,000–20,000 and so on to 90=>90,000
age	numeric	45–46	actual age 97=97 or older
political party	numeric	50–51	1=republican 3=democrat 5=independent 7=libertarian 9=other 11=no category
liberal-conservative	numeric	55	scale from 1 to 5 1=conservative 5=liberal

```
----+----1----+----2----+----3----+----4----+----5----+----6
1007 5  4  3 3 5  5   5  5    1  42   20   39   1  3
1009 1  5  5 5 3  1   9  3   13   .   35   69   1  1
1010 1  1  5 7 1  1   1  1   11   .   55   40   9  3
```

survresp.dat

An article by Fox, Crask, and Kim ("Mail Survey Response Rate," *Public Opinion Quarterly* (Winter 1988), Vol 52 No 4, 467–491) is a meta-analysis of experimental studies examining the impact of different incentives in mail surveys on response rates. These data (Table 2, pp. 480–481) look at the effect of the size of the incentive.

Number of observations = 30
Used in Modules 7, 9

variable	type	columns	additional description
incentive	numeric	1–4	values are $.10, $.25, $.50, and $1.00
sample size for control group	numeric	8–10	the control group did not receive the incentive
sample size for treatment group	numeric	13–15	the treatment group did receive the incentive
response rate for control group	numeric	18–22	the proportion of the control group who responded to the survey
response rate for treatment group	numeric	24–28	the proportion of the treatment group who responded

```
----+----1----+----2----+----3----+----4----+----5----+----6
0.25    150    150   0.247 0.373
0.25    150    150   0.193 0.380
0.50    100    100   0.540 0.680
```

taillite.dat

This data set contains a subset of data used in a study reported by McKnight and Shinar in "Brake Reaction Time to Center High-Mounted Stop Lamps on Vans and Trucks," *Human Factors* (Apr 1992), Vol 34 No 2, 205–213. This study investigated the benefits of using center high-mounted stop lamps (CHMSL), standard equipment on passenger cars since 1985, on trucks. The experiment was conducted by trapping a randomly selected driver between a test vehicle equipped with CHMSL and another vehicle that recorded the behavior of the test vehicle and the driver that followed. The dependent variable was response time, the time it took for the unsuspecting driver to brake after seeing the test vehicle brake. A. James McKnight provided the data and reported that as a result of this study, CHMSL is now standard equipment on trucks.

Number of observations =   1087
Used in Modules 11, 12, 13, 15, 16, 20, 21

variable	type	columns	additional description
ID	numeric	1-2	vehicle identification
vehicle type	numeric	4	1=pickup 2=cargo van 3=minivan 4=straight truck
group	numeric	7	1=experimental group (CHMSL on) 2=control group (CHMSL off)
position	numeric	10	CHMSL position 1=higher position 2=lower position
speed zone	numeric	13-14	in mph: 30, 40, or 50
response time	numeric	17-19	time from CHMSL/brake activation until subject vehicle braked
following time	numeric	23-24	lag time between test vehicle and subject vehicle in video frames
following time categories	numeric	28	based on following time 1=1-15 frames 2=16-30 frames 3=31-45 frames 4=46-60 frames

```
----+----1----+----2----+----3----+----4----+----5----+----6
1  1  1  2  30   18   14   1
2  1  1  2  30   28   29   2
3  1  2  2  30   27   19   2
```

utility.dat

This data set contains monthly records of phone, fuel, and electricity bills from August 1988 to February 1993.

Number of observations = 55
Used in Modules 2, 3, 5, 6, 7, 9, 10, 14

variable	type	columns	additional description
date	alphanumeric	1–6	
telephone costs	numeric	9–14	
fuel costs	numeric	17–22	
electricity costs	numeric	25–29	

```
----+----1----+----2----+----3----+----4----+----5----+----6
Aug 88  100.02   41.61  36.93
Sep 88   80.62   24.48  45.73
Oct 88   62.55   23.90  50.95
```

wear.dat

This data set was provided by Alex Kraftsov, a student in a design of experiments class. It measures the wear on a saw blade for different cut depths and grit sizes.

Number of observations = 36
Used in Module 13

variable	type	columns	additional description
grit size	alphanumeric	1–5	45/50 and 25/30
cut depth	numeric	8–9	25 and 75
wear	numeric	12–15	in m^2/mm

```
----+----1----+----2----+----3----+----4----+----5----+----6
45/50  25  20.6
45/50  25  33.3
45/50  25  17.6
```

well#1.dat

This data set is based on data collected by the Utah Division of Water Quality and provided by Dennis Frederick. Water quality is monitored by measuring particulate levels of several substances. These data came from one specific well for the period January 1990 to December 1991.

Number of observations = 33
Used in Modules 2, 4, 5, 6, 7, 10

variable	type	columns	additional description
date	alphanumeric	1–8	month/day/year
nitrate (mg/l)	numeric	11–14	
zinc (mg/l)	numeric	18–22	
TDS (mg/l)	numeric	25–27	total dissolved solids

```
----+----1----+----2----+----3----+----4----+----5----+----6
Jan09 90  0.76  0.004  893
Mar03 90  0.06  0.025  642
Apr20 90  0.06  0.005  524
```

well#8.dat

This data set is based on data collected by the Utah Division of Water Quality and provided by Dennis Frederick. Water quality is monitored by measuring particulate levels of several substances. This data set contains zinc levels for one specific well for the period January 1990 to October 1992.

Number of observations = 41
Used in Modules 7, 9

variable	type	columns	additional description
date	alphanumeric	1–8	month/day/year
zinc (mg/l)	numeric	11–15	

```
----+----1----+----2----+----3----+----4----+----5----+----6
Jan01 90  0.032
Mar02 90  0.277
Jul26 90  0.035
```

well#15.dat

This data set is based on data collected by the Utah Division of Water Quality and provided by Dennis Frederick. Water quality is monitored by measuring particulate levels of several substances. This data set contains total dissolved solids (TDS) levels for one specific well for the period January 1990 to December 1991.

Number of observations = 25
Used in Modules 7, 10

variable	type	columns	additional description
date	alphanumeric	1–8	month/day/year
TDS (mg/l)	numeric	11–14	total dissolved solids

```
----+----1----+----2----+----3----+----4----+----5----+----6
Jan01 91   1002
Jan23 91    974
Feb06 91    966
```

wine.dat

These data were collected by Craig Rasmussen, a student in a design of experiments class, as a class project. He wanted to know if temperature and brand affected the taste of wine. The response variable was a rating from 1 to 4, with 4 as the best. The experiment was conducted as a completely randomized design.

Number of observations = 36
Used in Modules 12, 20

variable	type	columns	additional description
wine brand	numeric	1	1=Glenn Ellen 3=Bel Arbors 2=Ernest and Julio Gallo
temperature	numeric	3	1=room temperature 2=slightly chilled 3=very cold
replicate	numeric	5	there are 4 reps
rating	numeric	7	1=poor 2=fair 3=good 4=excellent

```
----+----1----+----2----+----3----+----4----+----5----+----6
1 1 1 2
1 1 2 2
1 1 3 1
```

INDEX

regression • 107
    backward • 133
    confidence intervals • 107
    forward • 133
    PLOT statement • 109
    prediction intervals • 107
    stepwise • 133
repeated measures design • 96
RUN statement • 4

S

SAS variable _N_ • 25
SET statement • 20, 27
sorting by more than one variable • 18
spacing • 3
standard deviation • 40, 53
sums of squares, comparing types • 88

T

tables • 54
TABLES statement • 54
TITLE statement • 39
two-way tables • 141

V

variable names • 4
variables
    character • 5, 6
    length of character • 16
    numeric • 6